Selected Titles in This Series

(Continued in the back of this publication)

Existence and Persistence
of Invariant Manifolds for
Semiflows in Banach Space

MEMOIRS

of the
American Mathematical Society

Number 645

Existence and Persistence of Invariant Manifolds for Semiflows in Banach Space

Peter W. Bates
Kening Lu
Chongchun Zeng

September 1998 • Volume 135 • Number 645 (end of volume) • ISSN 0065-9266

American Mathematical Society
Providence, Rhode Island

1991 *Mathematics Subject Classification.*
Primary 58F15; Secondary 58F30, 58F35, 58G30, 58G35, 34C35.

Library of Congress Cataloging-in-Publication Data

Bates, Peter W., 1947–
  Existence and persistence of invariant manifolds for semiflows in Banach spaces / Peter W. Bates, Kening Lu, Chongchun Zeng.
    p. cm. — (Memoirs of the American Mathematical Society, ISSN 0065-9266 ; no. 645)
  "September 1998, volume 135, number 645 (end of volume)."
  Includes bibliographical references (p. –).
  ISBN 0-8218-0868-0 (acid-free paper)
  1. Differentiable dynamical systems. 2. Hyperbolic spaces. 3. Invariants. 4. Flows (Differentiable dynamical systems). I. Lu, Kening, 1962– . II. Zeng, Chongchun, 1973– . III. Title. IV. Series.
QA3.A57 no. 645
[QA614.9]
510 s—dc21
[515′.352]   98-25200
                                                                                            CIP

Memoirs of the American Mathematical Society

This journal is devoted entirely to research in pure and applied mathematics.

Subscription information. The 1998 subscription begins with volume 131 and consists of six mailings, each containing one or more numbers. Subscription prices for 1998 are $435 list, $348 institutional member. A late charge of 10% of the subscription price will be imposed on orders received from nonmembers after January 1 of the subscription year. Subscribers outside the United States and India must pay a postage surcharge of $30; subscribers in India must pay a postage surcharge of $43. Expedited delivery to destinations in North America $35; elsewhere $110. Each number may be ordered separately; *please specify number* when ordering an individual number. For prices and titles of recently released numbers, see the New Publications sections of the *Notices of the American Mathematical Society.*

  Back number information. For back issues see the *AMS Catalog of Publications.*

  Subscriptions and orders should be addressed to the American Mathematical Society, P. O. Box 5904, Boston, MA 02206-5904. *All orders must be accompanied by payment.* Other correspondence should be addressed to Box 6248, Providence, RI 02940-6248.

  Copying and reprinting. Individual readers of this publication, and nonprofit libraries acting for them, are permitted to make fair use of the material, such as to copy a chapter for use in teaching or research. Permission is granted to quote brief passages from this publication in reviews, provided the customary acknowledgment of the source is given.

  Republication, systematic copying, or multiple reproduction of any material in this publication (including abstracts) is permitted only under license from the American Mathematical Society. Requests for such permission should be addressed to the Assistant to the Publisher, American Mathematical Society, P. O. Box 6248, Providence, Rhode Island 02940-6248. Requests can also be made by e-mail to `reprint-permission@ams.org`.

Memoirs of the American Mathematical Society is published bimonthly (each volume consisting usually of more than one number) by the American Mathematical Society at 201 Charles Street, Providence, RI 02904-2294. Periodicals postage paid at Providence, RI. Postmaster: Send address changes to Memoirs, American Mathematical Society, P. O. Box 6248, Providence, RI 02940-6248.

CONTENTS

ABSTRACT

When considering the question of persistence of invariant manifolds for a dynamical system when that system is perturbed, one is naturally led to the concept of normal hyperbolicity. Roughly speaking, this requires that the tangent bundle of the phase space, restricted to the manifold, splits into three subbundles, one of which is the tangent bundle of the manifold and the linearized flow either expands or contracts the other bundles at rates greater than any expansion or contraction in the tangent bundle. Fenichel and Hirsch, Pugh and Shub proved that normal hyperbolicity is a sufficient condition for the persistence result in finite dimensional dynamical systems. Mañé proved that this condition is also necessary. Here we extend the sufficiency result to the case of semiflow in a Banach space. In the process of obtaining this we establish the existence of center-unstable and center-stable manifolds in a neighborhood of the unperturbed compact manifold.

1991 *Mathematics Subject Classification.* 58F15, 58F35, 58G30, 58G35, 34C35.

Key words and phrases. Normally hyperbolic invariant manifold, center-stable manifold, center-unstable manifold, semiflow.

The research of Peter Bates was supported in part by NSF grants DMS-9305044 and DMS-9622785.

The research of Kening Lu was supported in part by NSF grants DMS-9400233 and DMS-9622853.

Some of this work was completed while Peter Bates and Chongchun Zeng visited the Isaac Newton Institute for Mathematical Sciences, whose hospitality and support is gratefully acknowledged.

1. Introduction.

The general theory for dynamical systems has been developed by many mathematicians and scientists starting with Poincaré, Lyapunov and Birkhoff. The investigation of a particular dynamical system or a family of dynamical systems usually can be traced to an evolving physical system whose behavior one would like to understand and possibly predict. And so one of the main goal of the study of dynamical systems is to understand the long term behavior of states in the systems. In the original modeling process, empirical laws, simplifying assumptions, and even conjectured relationships are used to derive a dynamical system in the hope of then being able to approximately describe physical reality. Therefore, to have a better understanding of the physical phenomena being modeled, one needs to investigate not only the mathematical model but also the perturbations of the model. One also needs to study how the qualitative properties of the perturbed models are related to the qualitative properties of the original model. This is especially important in numerical computation. Because of round off error and imprecise numerical schemes, the model studied by numerical computations is actually a perturbation of the original model.

Although bifurcation theory, static and dynamic, are natural outcomes of this line of inquiry, for physical systems being modeled, one hopes that all the flows associated with small perturbations of the given system exhibit the same qualitative behaviors. When such is the case the dynamical system is said to be structurally stable, and the study of structural stability and its necessary or sufficient conditions has been a particularly fruitful field of study. If a dynamical system is not structurally stable, one may want to know when part of the qualitative properties are preserved under small perturbation. For instance, one may ask when equilibria perturb smoothly and what becomes of the flow in a neighborhood of an equilibrium as the system is perturbed. More generally, to understand the dynamics of a system, one needs to investigate the existence of invariant sets, in particular, such as equilibrium points, periodic orbits, invariant tori, and attractors, and then to study their structures and what happens in their vicinity. This leads to the fundamental problem of the persistence of invariant manifolds under perturbation and to the study of the qualitative properties of the flow near invariant manifolds. As outlined in "Historical Background" section below, this led to the establishment of the theory of normally hyperbolic invariant manifolds for finite dimensional dynamical systems, normal hyperbolicity (defined later) being exactly the right condition for persistence.

In this paper we initiate a program of extending the theory for normally hyperbolic invariant manifolds to infinite dimensional dynamical systems in a Banach space, thereby providing tools for the study of PDE's and other infinite dimensional equations of evolution.

Let X be a Banach space and let T^t be a C^1 semiflow defined on X, that is, it is continuous on $[0, \infty) \times X$, for each $t \geq 0, T^t : X \to X$ is C^1, and

$$T^t \circ T^s(x) = T^{t+s}(x)$$

Received by the editor October 27, 1995, and in revised form December 10, 1996.

for all $t, s \geq 0$ and $x \in X$. The typical example is the solution operator for a parabolic differential equation.

Suppose that there exists a smooth compact manifold, M, embedded in X which is invariant with respect to T^t, that is,

$$T^t(m) \in M \text{ for all } m \in M \text{ and } t \geq 0.$$

As examples we may think of critical points, periodic orbits or invariant tori, etc.

The questions which are addressed here concern the persistence of this invariant manifold under perturbations in the semiflow and the qualitative behavior of the semiflow near the invariant manifold. In general there will be no such manifold for the perturbed flow even in finite dimensional space, see examples given by Hale [Hal3]. However, if there is a certain nondegeneracy of the linearized flow at all points of M then one expects even smooth continuation of M with sufficiently small perturbations of T^t. This is as in the Implicit Function Theorem which provides smooth continuation of a zero of a function under smooth perturbation provided the linearization of the function at the zero is nonsingular. For an invariant *manifold*, rather than just a critical point, the nondegeneracy condition is *normal hyperbolicity*. This condition gives, for each $m \in M$, a decomposition $X = X_m^c \oplus X_m^u \oplus X_m^s$, with X_m^c the tangent space to M at m such that

(a) This splitting is invariant under the linearized flow, $DT^t(m)$ and

(b) $DT^t(m)|_{X_m^u}$ expands and does so to a greater degree than does $DT^t(m)|_{X_m^c}$ while $DT^t(m)|_{X_m^s}$ contracts and does so to a greater degree than does $DT^t(m)|_{X_m^c}$.

The superscripts c, u and s stand for "center," "unstable, " and "stable." The precise definition of normal hyperbolicity is given in Section 2, where we give notation and collect some fundamental preliminaries.

Our main results may be summarized as

Theorem. *Suppose that T^t is a C^1 semiflow on X and M is a C^2 compact connected invariant manifold on which T^t is normally hyperbolic for t sufficiently large. Suppose also that for each $m \in M, DT^t|_{X_m^u}$ is an isomorphism. Let $t_1 > 0$ be large enough and be fixed and N be a fixed neighborhood of M. For any $\epsilon > 0$ there exists $\sigma > 0$ such that if $\tilde{T}$ is a C^1 map on X which satisfies $\|\tilde{T} - T^{t_1}\|_{C^1(N)} < \sigma$, then*

(a) *Persistence: $\tilde{T}$ has a unique C^1 compact connected normally hyperbolic invariant manifold $\tilde{M}$ near M.*

(b) *Convergence: $\tilde{M}$ converges to M in the C^1 topology as $\|\tilde{T} - T^{t_1}\|_{C^1}$ tends zero;*

(c) *Existence: $\tilde{T}$ has unique C^1 invariant manifolds $\tilde{W}^{cs}(\epsilon)$ and $\tilde{W}^{cu}(\epsilon)$ in a tubular neighborhood $N(\epsilon)$ of M, which at M are tangent to the center-stable vector bundle $\tilde{X}^c \oplus \tilde{X}^s$ and the center-unstable vector bundle $\tilde{X}^c \oplus \tilde{X}^u$, respectively.*

(d) *Characterization:*

$$\tilde{W}^{cs}(\epsilon) = \left\{ x_0 \in N(\epsilon) : \tilde{T}^k(x_0) \in N(\epsilon), \text{ for } k \geq 1, \tilde{T}^k(x_0) \to \tilde{M} \text{ as } k \to \infty \right\},$$

$$\tilde{W}^{cu}(\epsilon) = \left\{ x_0 \in N(\epsilon) : \exists \{x_k\}_{k>0} \subset N(\epsilon), \text{ satisfying} \right.$$

$$\left. \tilde{T}(x_k) = x_{k-1} \text{ for } k \geq 1, \text{ and } x_k \to \tilde{M} \text{ as } k \to \infty \right\}$$

and $\tilde{M} = \tilde{W}^{cs}(\epsilon) \cap \tilde{W}^{cu}(\epsilon)$.

Furthermore, if $\tilde{T}^t$ is a C^1 semiflow on X which satisfies $\|\tilde{T}^{t_1} - T^{t_1}\|_{C^1(N)} < \sigma$ and $\|\tilde{T}^t - T^t\|_{C^0(N)} < \sigma$ for $0 \leq t \leq t_1$, then $\tilde{M}$ is a normally hyperbolic invariant manifold for $\tilde{T}^t$ with center-stable manifold $\tilde{W}^{cs}(\epsilon)$ and center-unstable manifold $\tilde{W}^{cu}(\epsilon)$, respectively.

Our Theorems are stated precisely in Section 3.

Remark. From property (d) we see that the local center-stable manifold $\tilde{W}^{cs}$ consists of points for which all forward iterates lie in the tubular neighborhood and approach $\tilde{M}$. Hence, $\tilde{W}^{cs}$ can be regarded as the stable manifold of $\tilde{M}$. Likewise, the local center-unstable manifold, $\tilde{W}^{cu}(\epsilon)$, consists of points for which backward orbits exist, stay in the tubular neighborhood for all backward iterates, and approach $\tilde{M}$. Hence, this can be regarded as the unstable manifold of $\tilde{M}$.

It is instructive to see what this theorem says in simple settings, where the T^t is the solution operator for a linear differential equation

$$x' = Ax, \qquad x \in \mathbb{R}^n$$

where A is an $n \times n$ matrix of which all eigenvalues have either positive real part or negative real part. In this case, the equilibrium point $x = 0$ is normally hyperbolic, which is the same as the concept of hyperbolicity. The eigenspace corresponding the eigenvalues with negative real part is the stable space (manifold) of the equilibrium 0 and the eigenspace corresponding the eigenvalues with positive real part is the unstable space (manifold) of the equilibrium 0. For a small C^1 perturbation $f(x)$,

$$x' = Ax + f(x)$$

has a unique hyperbolic equilibrium point $\tilde{x}$ near 0 by the Implicit Function Theorem and has also stable and unstable manifolds at $\tilde{x}$ which are the perturbations of the stable space and the unstable space at 0, respectively. These invariant manifolds play a key role in the study of the structural stability of the flow near hyperbolic equilibrium points (see [Har] for instance).

A more complicated example is a periodic orbit of an ordinary differential equation

$$x' = f(x), \qquad x \in \mathbb{R}^n.$$

Let $p(t)$ be the periodic solution with period L and let $\Phi(t)$ be the fundamental matrix of

$$x' = Df(p(t))x.$$

The periodic orbit $M = \{p(t) : 0 \leq t \leq L\}$ is normally hyperbolic if all eigenvalues of the time-L map $\Phi(L)$ are off the unit circle except 1, which is simple.

Historical Background.

The theory of invariant manifolds for discrete and continuous dynamical systems has a long and rich history.

For the case of M consisting of a single equilibrium point, Hadamard [Ha] constructed the unstable manifold of a hyperbolic fixed point of a diffeomorphism of the plane by iterating the mapping applied to a curve in the plane, thereby obtaining a convergent sequence of curves. The limit of this sequence of curves gives the unstable manifold. This geometric method is now called Hadamard's graph transform.

Lyapunov [Ly] and Perron [P1], [P2], [P3] constructed the unstable manifold of an equilibrium point by formulating the problem in terms of an integral equation. This method is analytic rather than geometric and now is called the method of Lyapunov-Perron. Although the successful application of this approach tends to give more information about the smoothness of the manifold, certain obstacles which we explain later, lead us to use Hadamard's approach.

There is an extensive literature on the stable, unstable, center, center-stable, and center-unstable manifolds of equilibrium points for both finite and infinite dimensional dynamical systems. We do not attempt to give an exhaustive list of references. The general theory for finite dimensional dynamical systems may be found in [BDL], [Ca], [HP], [HPS], [Ir], [Ke], [Ku], [MS], [Pl], [Si], [Sm], [Va], and [VV]. For infinite dimensional dynamical systems we refer the reader to [Ba], [BJ], [CL1], [He], [Mi], [VI, and [Wa]. Most of these works use the approach of Lyapunov- Perron. A good treatment of center manifold theory using the Lyapunov-Perron approach for the finite dimensional case may be found in the monograph by Carr [Ca], where many applications are also set forth. Certain infinite dimensional settings are also treated. Vanderbauwhede and Van Gils [VV] also use the Lyapunov Perron method to obtain smooth center manifolds but with some important differences in technique. Ball [Ba] used the Lyapunov-Perron approach to obtain local stable, unstable and center manifolds for equilibrium points of dynamical systems in Banach space, with application to the beam equation. Henry [He] developed the theory for semilinear parabolic equations. Later, Chow and Lu [CL1] used this approach to prove the existence of smooth center-unstable manifolds with application to the damped wave equation. For more on center manifold theory in the infinite dimensional setting, using the Lyapunov-Perron method, see [VI]. The theory of invariant manifolds for an equilibrium point of finite dimensional dynamical using Hadamard's approach may be found in [HP]. For infinite dimensional dynamical systems, we refer to [BJ], where applications are given demonstrating the stability of a pulse solution to the FitzHugh-Nagumo equations and the instability of stationary solutions to the nonlinear Klein-Gordon equation.

Krylov and Bogoliubov [KB] studied time-periodic ordinary differential equations arising from the study of nonlinear oscillations. Under the assumption that the averaged equation has an asymptotically stable equilibrium point, they proved the existence of periodic integral manifolds, which gives the existence of asymptotically stable periodic orbits for a class of equations. An integral manifold is an invariant manifold in the product space of time and phase space. The above result and many generalizations and related work is collected in the monograph of Bigoliubov and Mitropolsky [BM].

Levinson [Le] studied periodic perturbations of an autonomous ordinary differential equation possessing an asymptotically stable periodic orbit. He proved that if the perturbation was sufficiently small, then the perturbed equation has a periodic integral manifold, which may be viewed as a two-dimensional torus. Levinson's results were extended to periodic surfaces by Diliberto [Di], Hufford [Hu], Marcus [Ma], and Kyner [Ky1].

Hale [Ha1] established a general theory of integral manifolds for nonautonomous ordinary differential equations and obtained more general results than those just mentioned above. An extension of Hale's integral manifold theory to a larger class of nonautonomous ordinary differential equations was recently obtained in [Yi].

McCarthy [Mc] studied the persistence of a compact normally hyperbolic invariant manifold for a diffeomorphism, where the normal bundle is the stable normal bundle. McCarthy's work was extended by Kyner [Ky2], Kurzweil [Ku] and Nyemark [Ny]. Sacker gave further extensions in [Sa].

The most general theory of compact normally hyperbolic invariant manifolds for finite dimensional dynamical systems were independently obtained by Hirsch, Pugh and Shub [HPS1], [HPS2] and Fenichel [F1-3]. They proved the persistence of normally hyperbolic invariant manifolds, and the existence of the center-stable and center-unstable manifolds and their invariant foliations.

Mañé [Mn1] proved that normal hyperbolicity defined in [HPS] is a necessary condition for the persistence of an invariant manifold for finite dimensional dynamical systems.

Recently, Pliss and Sell [PS] studied the persistence of hyperbolic attractors for ordinary differential equations. Sell [Se] has reported that their methods also extend to the infinite dimensional setting.

Henry [He] extended Hale's theory of integral manifolds to the most general nonautonomous abstract semilinear parabolic equations. His result can also be applied to ordinary differential equations. Henry also studied compact normally hyperbolic invariant manifolds with trivial normal bundle for semilinear parabolic equations and obtained a coordinate transformation which leads to the setting of integral manifolds. His results suggest how one may obtain the persistence of compact normally hyperbolic invariant manifolds under perturbation, by employing the Lyapunov-Perron approach.

There has been much recent work surrounding one class of global manifolds, namely inertial manifolds, which are global, attracting, finite dimensional manifolds for dissipative systems. As such these manifolds are rather robust objects which are large enough to contain most of the interesting dynamics of a system. That they are finite dimensional means a substantial reduction in phase space is possible when one is interested in the asymptotic behavior of solutions. A general theory may be found in

[FST], [M-PS] and [CFNT]. Our results can not be directly applied to the persistence of inertial manifolds. However, our analysis may be applied to obtain results along these lines.

Very recently, Li, McLaughlin, Shatah and Wiggins [LMSW] have proved the persistence of the center-stable and center-unstable manifolds of a circle of stationary solutions for the nonlinear Schrödinger equation as it is perturbed. Note that for the NLS equation the solution operator is a group. Linearizing at this particular circle of stationary solutions, which happens to consist of spatially constant functions, gives trivial stable and unstable bundles. Therefore a global change of variables is possible in a neighborhood of this circle, distinguishing center, stable, and unstable directions. By using the Lyapunov-Perron approach, they proved persistence of center-stable and center-unstable manifolds of the circle. They also produce invariant foliations of center-stable (with stable leaves) and center-unstable (with unstable leaves) manifolds of perturbations of the circle. These foliations are very useful in tracking trajectories and completing a shooting argument to discover homoclinic orbits. In [LMW], Li, McLaughlin, and Wiggins studied the persistence of overflowing manifolds of finite codimension for a C^r ($r \geq 2$) group S^t in a Hilbert space. In this case, for each fixed $t \in \mathbb{R}$, S^t is a C^r diffeomorphism. Under certain assumptions such as trivial normal bundle, uniform boundedness of the second derivative, $D^2 S^t$, and that the overflowing manifold is covered by finitely many balls, they obtained the persistence of the overflowing manifold and the existence of an invariant foliation of the overflowing manifold. They did this by using the method of Hadamard's graph transform. They also applied these results to the Schrödinger equation to recover the results of [LMSW].

The versions of the theorems we will establish, in the case that the time-t map is a diffeomorphism, as in the finite dimensional setting, were proved by Hirsch, Pugh and Shub [HPS]. Simultaneously and independently, N. Fenichel [F1] proved a similar result with a somewhat different definition of normal hyperbolicity in terms of a variant of Lyapunov number. Our definition of normal hyperbolicity, given in Section 2, is modeled on that given in [HPS]. In both [F1] and [HPS] Hadamard's graph transform is used to develop the perturbation theory for normally hyperbolic compact invariant manifolds. Their proofs are very elaborate with many details left to the reader. Recently, Wiggins [W] provided more detailed proofs of Fenichel's theorems while Bronstein and Kopanskii [BK] provided the detailed proof of Hirsch, Pugh and Shub's theorem on the persistence of normally hyperbolic invariant manifolds.

There are several differences between a finite dimensional dynamical system and an infinite dimensional dynamical system. The most significant difference is that the infinite dimensional dynamical systems generated, for example, by parabolic equations, define only semiflows since backwards solutions may not exist. Thus the time-one maps associated with these dynamical systems are not invertible. Furthermore, the phase spaces are not locally compact. In addition, the trivialization of the normal bundle of a manifold, the local orthogonal coordinates, and some results in differential geometry such as the results on smoothing of bundles are not available in Banach spaces. These make the study of normally hyperbolic invariant manifolds for a semiflow in a Banach space much more complicated. In our analysis, the lack of local compactness is mitigated by the compactness of the original manifold. However, the lack of inverse and global trivialization were significant obstacles which greatly added to length of this paper.

The results given here lay the groundwork for constructing invariant foliations in a neighborhood of a compact normally hyperbolic invariant manifold for a semiflow. Invariant foliations for flows or semiflows are extremely useful in that they can be used to track the asymptotic behavior of solutions and to provide coordinates in which systems of differential equations may be decoupled and normal forms derived. In [HS] and in Fenichel's work [F2], [F3], invariant foliations are obtained and some of their uses demonstrated. Since then, the applications of this theory to problems from science and engineering have flourished, especially applications to singular perturbation problems, see for example [D], [G], [GS], [HW], [JK], [KW], [Li], [Sz] and [Te]. Recently Jones [Jo] has given a clear discussion of the use of Fenichel's theorems as they apply to singular perturbation problems. He includes proofs of these theorems and important extensions of the λ-lemma (see also [JK]). Kirchgraber and Palmer [KP] have recently given detailed results on invariant foliations and their applications to linearizations for finite dimensional systems.

Ruelle [Ru] proved a result giving invariant stable and unstable fibrations almost everywhere on a compact invariant set for a semiflow in Hilbert space. It was assumed that the linearized time-t map is compact and injective with dense range. The results are therefore applicable to some parabolic PDE's. Mañé [Mn2] extended Ruelle's results to semiflows in Banach space.

Considering the semiflow generated by a parabolic equation, Lu in [Lu] constructed infinitely many invariant manifolds as perturbations of eigenspaces of the operator obtained by linearizing at an equilibrium point. With these and corresponding invariant foliations, a new coordinate system was constructed in a neighborhood of the equilibrium point. This facilitated a proof of a Hartman-Grobman theorem for scalar parabolic PDE's, which yields that the flow near a hyperbolic equilibrium point is structurally stable. In [BL] a more general theorem on the existence of invariant foliations was proved. This theorem was then used to obtain a Hartman-Grobman result for both the phase-field system and the Cahn-Hilliard equation. The previous two papers are concerned with dynamics in a neighborhood of an equilibrium but Chow, Lin and Lu [CLL] proved a general result giving a stable fiber at each point of an inertial manifold, thereby giving a more global invariant foliation of infinite dimensional space.

Recently, Aulbach and Garay [AG] used invariant foliations to study partial linearization for noninvertible mappings near fixed points.

In the present paper we deal only with the persistence of a normally hyperbolic invariant manifold and the existence of the center-stable and center-unstable manifolds, leaving invariant foliations to a forthcoming paper [BLZ1]. However, this paper sets the stage for that development. We shall discuss application of the results obtained in this paper and in [BLZ1] elsewhere.

Nontechnical Overview.

Our proof for the existence and persistence of invariant manifolds is based on Hadamard's graph transform and consists of seven main steps:

Step 1. Coordinate Systems.

We first build a C^1 tubular neighborhood of M. Then we introduce three coordinate systems. The first coordinate system is given by the tubular neighborhood, identifying it with a neighborhood, $N(\epsilon)$, in the normal bundle for the manifold. The second is defined by the splitting of the tangent bundle of the phase space X restricted to M, which is a Cartesian system in a neighborhood of a point on M. The third coordinate system is induced by the local trivialization of the bundles based on M. We then estimate how the coordinates differ from each other. These details are given in Section 4. The main reason for these coordinate systems is that we will be looking at Lipschitz graphs over the unstable (or stable) bundle of M. The graph is a global geometric object but the Lipschitz property requires local coordinates for its description. The local trivialization relates two nearby local coordinate systems.

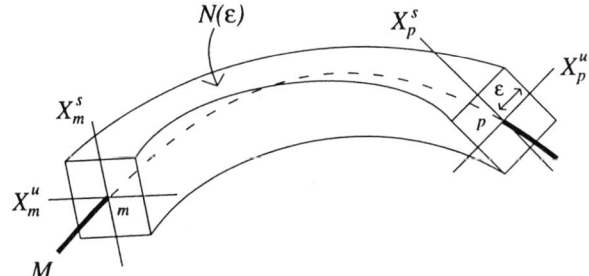

Figure 1. a) Tubular Neighborhood

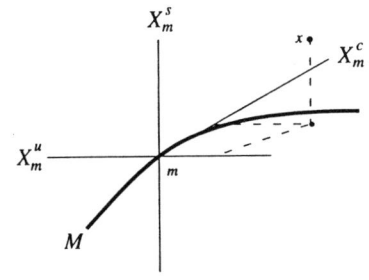

$$x = m + x^c + x^u + x^s, \qquad x^\alpha \in X_m^\alpha$$

Figure 1. b) Local Cartesian Coordinates

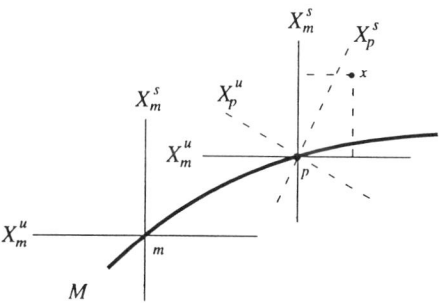

$$x = p + x^u + x^s = p + \tilde{x}^u + \tilde{x}^s, \quad x^\alpha \in X_p^\alpha, \quad \tilde{x}^\alpha \in X_m^\alpha$$

Figure 1 c) Local Trivialization

Step 2. Invariant Cones.

In Section 2, we define various cones: The global cone with 'vertex' M and 'axes' being the stable and unstable bundles, the global fattened cone which includes a small neighborhood of the 'vertex', and the global gap cone in which a small neighborhood of the 'vertex' is deleted (see Figure 2). By using the normal hyperbolicity of M, we show that these cones are invariant for the unperturbed time$-t_1$ map, T^{t_1}. Furthermore, the global fattened cone and the global gap cone are invariant for a small C^1 perturbation $\tilde{T}$ of T^{t_1}. These statements are proved in Section 5. The invariance of local cones moving with the semiflow is also proved in Section 5. The invariance of these cones play crucial roles in establishing the existence of invariant manifolds.

Step 3. Existence of Lipschitz Center-Unstable Manifold.

We regard the normal bundle $X^u \oplus X^s$ as a bundle over the unstable bundle X^u and consider sections of this bundle in the tubular neighborhood $N(\epsilon)$. Define the complete metric space Γ^{cu} of Lipschitz sections, which are described in terms of the global fattened cone and local cones. The fattened cone (and the gap cone) are introduced to allow for perturbations. The key to construct a "graph transform" in the space Γ^{cu} is to show that for any given Lipschitz section $h \in \Gamma^{cu}$, the image of the graph of this section under $\tilde{T}$ restricted to the tubular neighborhood is the graph of a section in Γ^{cu}. This yields a graph transform $\mathcal{F}^{cu}$ defined in Γ^{cu}. Using the invariance of the cones, one can prove that $\mathcal{F}^{cu}$ is a contraction and the fixed point is the desired center-unstable manifold. This is done in Section 6, where some properties of the center-unstable manifold are also discussed.

Step 4. Existence of Lipschitz Center-Stable Manifold.

For a finite dimensional dynamical system, one may obtain the stable manifold as the unstable manifold for the corresponding time reversed system(see, for example, [HPS]). However, this technique does not work for infinite dimensional dynamical systems, for example those generated by parabolic equations, since backward solutions may not exist and the corresponding solution map is not invertible. In order to overcome this difficulty, a new technique is needed to establish the existence of the center-stable manifold. We again introduce a complete metric space Γ^{cs} of Lipschitz sections, this time, of the bundle $X^u(\epsilon) \oplus X^s(\epsilon)$ over $X^s(\epsilon)$. For each Lipschitz section $h \in \Gamma^{cs}$, since $\tilde{T}$ may not be invertible, a point on the graph of h may have no preimage or have more than one preimage. Thus the preimage of the graph of the section h under $\tilde{T}$ can be very complicated. In Section 7, we find a way to construct a unique Lipschitz section in Γ^{cs} whose the graph is perhaps just a part of the preimage of the graph of h under $\tilde{T}$. This is the key idea, this construction inducing a graph transform in Γ^s. Finally we show that the this graph transform is a contraction and the fixed point is the desired stable manifold. Again the invariance of the cones is used.

Step 5. The Smoothness of the Center-Unstable and Center-Stable Manifolds.

For the center unstable manifold, the basic idea to show its smoothness is to find a candidate for the tangent bundle of this manifold, which is invariant under the linearization $D\tilde{T}$, then to prove it indeed is tangent to the manifold. The arguments are based on the use of Lipschitz jets, which is borrowed from [HPS]. Since the trivialization of the normal bundle is not available in a Banach space, the proof is more complicated than for finite dimensional systems. We first define a space of sections of the Lipschitz jet bundle, which is different from the jet spaces introduced in [HPS]. Then we construct a graph transform based on the linearization $D\tilde{T}$ and show that it has a unique fixed point which gives the tangent bundle of the center-unstable manifold. A major difficulty in finding the fixed point is that the space of sections of the Lipschitz jet bundle is not complete. Finally, we prove that the tangent bundle is C^0. For the center-stable manifold, an additional difficulty is that $D\tilde{T}$ may not be invertible. One can use similar arguments to those which gave the existence of the center-stable manifold to construct the graph transform in the space of sections of the Lipschitz jet bundle.

Step 6. The Normal Hyperbolicity

In Section 10, we show that the intersection of the center-stable manifold and center-unstable manifold is a C^1 compact connected invariant manifold. The basic idea to obtain normal hyperbolicity for the perturbed manifold $\tilde{M}$ is to construct stable and unstable bundles from the tangent bundles of the center-stable and center-unstable manifolds by finding projection operators. This is done in Section 11. Summarizing all results obtained so far gives the results for the perturbed *map* $\tilde{T}$.

Step 7. Results for the Perturbed Semiflow

In Section 12 we show that the results for maps which are C^1-close to the map T^{t_1} also hold for semiflows whose time-t_1 maps are C^1-close to T^{t_1} and whose time$-t$ maps are C^0 close to T^t for $t \in [0, t_1]$.

Acknowledgement. We would like to acknowledge several fruitful conversations with Shui-Nee Chow, Jack Hale, Chris Jones, and Xiao-Biao Lin. PWB and CZ would like to thank the faculty and staff of the Isaac Newton Institute for their hospitality and support during the final stages of this work. We also would like to thank the referee for valuable suggestions.

2. Notation and Preliminaries.

The results and proofs presented in this paper require a certain amount of technical notation which we collect in this section for future reference. Throughout, X will represent a Banach space with norm $|\cdot|$. In subspaces the same norm symbol is used. The notation $\|\cdot\|$ will be reserved for the linear operator norm $\|L\| \equiv \sup\{|Lx| : |x| = 1\}$.

When nonlinear operators are considered on a bounded subset $B \subset X$ we use the notation

$$\|F\|_0 = \sup\{|F(x)| : x \in B\}$$

and for $k \geq 1$

$$\|F\|_k = \|F\|_0 + \sum_{j=1}^{k} \sup\{\|D^j F(x)\| : x \in B\},$$

expecting the choice of B to be clear from the context. Here, D is the derivative operator.

The open ball centered at $x \in X$ of radius $r > 0$ will be denoted by $B(x, r)$.

On X we suppose that there is a semiflow, T^t, for which the following conditions hold:

(H1) it is continuous on $[0, \infty) \times X$ into X, and for each $t \geq 0, T^t : X \to X$ is C^1.

(H2) There exists a C^2 compact connected invariant manifold $M \subset X$, i.e., $T^t(M) \subset M$ for each $t \geq 0$.

(H3) M is **normally hyperbolic**, that is
  (i) for each $m \in M$ there is a decomposition

$$X = X_m^c \oplus X_m^u \oplus X_m^s$$

  of closed subspaces with X_m^c the tangent space to M at m.
  (ii) For each $m \in M$ and $t \geq 0$, if $m_1 = T^t(m)$

$$DT^t(m)\big|_{X_m^\alpha} : X_m^\alpha \to X_{m_1}^\alpha \qquad \text{for } \alpha = c, u, s$$

  and $DT^t(m)|_{X_m^u}$ is an isomorphism from X_m^u onto $X_{m_1}^u$.
  (iii) There exists $t_0 \geq 0$ and $\lambda < 1$ such that for all $t \geq t_0$

$$\lambda \inf\left\{\left|DT^t(m)x^u\right| : x^u \in X_m^u, |x^u| = 1\right\} > \max\left\{1, \left\|DT^t(m)\big|_{X_m^c}\right\|\right\} \qquad (2.1)$$

$$\lambda \min\left\{1, \inf\left\{\left|DT^t(m)x^c\right| : x^c \in X_m^c, |x^c| = 1\right\}\right\} > \left\|DT^t(m)\big|_{X_m^s}\right\| \qquad (2.2)$$

Condition (2.1) suggests that near $m \in M$, T^t is expansive in the direction of X_m^u, and at a rate greater than that on M, while (2.2) suggests that T^t is contractive in the direction of X_m^s, and at a rate greater than that on M. Thus, fibers X_m^c, X_m^u and X_m^s are distinguished by the growth and decay rates of the flow, much as one sees with an exponential trichotomy. A consequence of (H3) is the following

Lemma 2.1. *For all $t \geq 0$, T^t is a C^1 diffeomorphism on M.*

Proof. Suppose that for some $t_1 > 0$ and points of $M, m_1 \neq m_2$, $T^{t_1}(m_1) = T^{t_1}(m_2)$. Let $\bar{t} = \inf\{t > 0 : T^t(m_1) = T^t(m_2)\}$ and denote $T^{\bar{t}}(m_1) = T^{\bar{t}}(m_2)$ by $\bar{m}$.

Note that for $t_2 > t_0$, $DT^{t_2}(m)\big|_{X_{\bar{m}}^c}$ is invertible by (iii), and so, by the Inverse Function Theorem, there exist neighborhoods U_1 of $\bar{m}$ and U_2 of $T^{t_2}(\bar{m})$ such that T^{t_2} is a diffeomorphism from $U_1 \cap M$ onto $U_2 \cap M$. Now take $\delta \in (0, t_2)$ so small that $T^{\bar{t}-\delta}(m_1), T^{\bar{t}-\delta}(m_2) \in U_1 \cap M$, then $T^{t_2}\left(T^{\bar{t}-\delta}(m_1)\right) = T^{t_2-\delta}\left(T^{\bar{t}}(m_1)\right) = T^{t_2-\delta}(\bar{m}) = T^{t_2}\left(T^{\bar{t}-\delta}(m_2)\right)$, contradicting the fact that T^{t_2} is one-to-one on $U_1 \cap M$. Therefore, for $t_1 > 0$, $T^{t_1}|_M$ is one-to-one.

Again, using the fact that for $t \geq t_0$ $DT^t|_{X^c}$ is an isomorphism, $T^t(M)$ is open in M, by the Inverse Function Theorem. Since M is compact, $T^t(M)$ is closed also, and so $T^t(M) = M$. Consequently, for any $t \geq t_0$, $T^t|_M$ is a diffeomorphism. For $t \in [0, t_0]$ $T^t \circ T^{t_0-t}|_M = T^{t_0-t} \circ T^t|_M = T^{t_0}|_M$, which is a diffeomorphism, and therefore T^t is also a diffeomorphism. $\square$

Similarly, $DT^t(m)\big|_{X_m^c}$ is invertible with inverse uniformly bounded for t in compact intervals. we have

Lemma 2.2. *For any $t_1 > 0$, there exists a constant C_1 such that for all $m \in M$ and $0 \leq t \leq t_1$*

$$C_1^{-1} \leq \inf\left\{|DT^t(m)x^c| : x^c \in X_m^c, |x^c| = 1\right\} \leq \|DT^t(m)|_{X_m^c}\| \leq C_1.$$

Proof. Let $M_d^2 = \{(m_1, m_2) : m_1, m_2 \in M, m_1 \neq m_2\}$. For $i = 1, 2$, we define functions f_i from $[0, t_1] \times M_d^2$ to $\mathbb{R}^+$ by

$$f_1(t, m_1, m_2) = \frac{|T^t(m_1) - T^t(m_2)|}{|m_1 - m_2|},$$

$$f_2(t, m_1, m_2) = \frac{|m_1 - m_2|}{|T^t(m_1) - T^t(m_2)|}.$$

For $m_1, m_2 \in M$, define

$$d(m_1, m_2) = \inf\left\{\int_0^1 |\gamma'(\tau)|d\tau : \gamma \in C^1([0, 1], M), \gamma(0) = m_1, \gamma(1) = m_2\right\}.$$

It is easy to show that $d(\cdot, \cdot)$ is a metric on M, and induces an equivalent topology for M. In fact, we first note that $d(m_1, m_2) \geq |m_1 - m_2|$, then by the compactness of M there exits a constant C such that $d(m_1, m_2) \leq C|m_1 - m_2|$. For each fixed $0 \leq t \leq t_1$, from Lemma 2.1,

$$\max_{m \in M} \|DT^t(m)|_{X_m^c}\| d(m_1, m_2) \geq d(T^t(m_1), T^t(m_2))$$

$$\geq \left(\max_{m \in M} \|D(T^t|_M)^{-1}(m)\|\right)^{-1} d(m_1, m_2),$$

which implies that $f_1(t, \cdot, \cdot)$ and $f_2(t, \cdot, \cdot)$ are bounded for each fixed $0 \le t \le t_1$. Let

$$E_k = \{t \in [0, t_1] : f_i(t, m_1, m_2) \le k, i = 1, 2, (m_1, m_2) \in M_d^2\}.$$

Clearly, $[0, t_1] = \cup_1^\infty E_k$. By the Baire Category Theorem, there exits $0 \le t' < t'' \le t_1$ and $k > 0$ such that for all $t \in [t', t'']$ and $(m_1, m_2) \in M_d^2$

$$f_1(t, m_1, m_2) \le k \text{ and } f_2(t, m_1, m_2) \le k,$$

which implies

$$\|DT^t(m)|_{X_m^c}\| \le k \text{ and } \|D(T^t|_M)^{-1}(m)|_{X_m^c}\| \le k.$$

Thus, for $t \in [0, t'' - t']$ and $m \in M$,

$$\|DT^t(m)|_{X_m^c}\| = \|DT^{t+t'}\left((T^{t'}|_M)^{-1}(m)\right)D(T^{t'}|_M)^{-1}(m)|_{X_m^c}\| \le k^2.$$

Similarly,

$$\|D(T^t|_M)^{-1}(m)|_{X_m^c}\| \le k^2.$$

Therefore, by using the semigroup property and the fact that $[0, t_1]$ is compact, there exists a constant C_1 such that Lemma 2.2 holds. This completes the proof. $\square$

One may ask if (H3) (iii) is need for all $t \ge t_0$. In fact, we have the following

Lemma 2.3. *Let T and M satisfy (H1), (H2) and (i), (ii) in (H3). If there exists $0 < t_0 < t_1$ such that (2.1) and (2.2) hold for all $t \in [t_0, t_1]$, then (2.1) and (2.2) also hold for all $t \ge (1 + [\frac{t_0}{t_1 - t_0}])t_0$, where $[\frac{t_0}{t_1 - t_0}]$ denotes the integer part of the number $\frac{t_0}{t_1 - t_0}$.*

Proof. Let k be a positive integer. For $t \in [kt_0, kt_1]$ and $m \in M$, let $m_i = T^{i\frac{t}{k}}(m)$ for $i = 0, 1, 2, \cdots, k - 1$. Observe $t_0 \le t/k \le t_1$ and

$$DT^t(m) = DT^{\frac{t}{k}}(m_{k-1})DT^{\frac{t}{k}}(m_{k-2}) \cdots DT^{\frac{t}{k}}(m_0).$$

Thus, by the assumption of this lemma, we have that (2.1) and (2.2) hold for all $t \in [kt_0, kt_1]$. On the other hand, for $k \ge 1 + [\frac{t_0}{t_1 - t_0}]$,

$$kt_1 - (k + 1)t_0 = k(t_1 - t_0) - t_0 \ge (1 + [\frac{t_0}{t_1 - t_0}])(t_1 - t_0) - t_0 \ge 0$$

which implies

$$[(1 + [\frac{t_0}{t_1 - t_0}])t_0, \infty) \subset \cup_{k=1}^\infty [kt_0, kt_1].$$

This completes the proof. $\square$

Associated with the decomposition of X introduced in (H3) are vector bundles

$$X^\alpha \equiv \{(m, X_m^\alpha) : m \in M\}, \ \alpha = c, u, \text{ and } s.$$

In the usual way we define the direct sum of bundles, e.g.,

$$X^u \oplus X^s \equiv \{(m, x^u + x^s) : x^u \in X_m^u \text{ and } x^s \in X_m^s, m \in M\}.$$

For $\epsilon > 0$ we shall use the notation

$$X^\alpha(\epsilon) \equiv \{(m, x^\alpha) : x^\alpha \in X_m^\alpha, m \in M, |x^\alpha| < \epsilon\}, \alpha = c, u, \text{ and } s,$$

and

$$X^u(\epsilon) \oplus X^s(\epsilon) \equiv \{(m, x^u + x^s) : x^u \in X_m^u, x^s \in X_m^s, m \in M, |x^u| < \epsilon, |x^s| < \epsilon\},$$

the latter of which may be identified with a tubular neighborhood of M, as will be seen in Section 4. The mapping which does this is $\Theta : X^u \oplus X^s \to X$ defined by $\Theta(m, x^u + x^s) = m + x^u + x^s$. For $m \in M$ we shall also use projections, Π_m^α, associated with the decomposition of X given in (H3), for $\alpha = c, u$, and s. We assume

(H4) The mapping from $M \subset X \to \mathcal{L}(X)$, the continuous linear operators on X, defined by $m \to \Pi_m^\alpha$ is C^1 for $\alpha = c, u$, and s.

In [BLZ2], we remove the assumption on the smoothness of the bundles, that is, (H4) can be removed and the assumption that the manifold M is C^2 can be relaxed to require only C^1.

Remark. If $|\cdot|_{X^\alpha}$ is defined by $|(m, x)|_{X^\alpha} = |x|$ with $|\cdot|$ the norm in X, then the continuity in m of the above mapping gives the bundle X^α a so-called Finsler structure.

To show that Θ is a diffeomorphism we will require neighborhoods to be so small that Π_m^α changes very little as m varies in the neighborhood. We therefore give the following:

Definition: For $m_0 \in M$ and $\eta \in (0, 1)$, a neighborhood U of m_0 in X is said to be an η-neighborhood if $\|\Pi_m^\alpha - \Pi_{m_0}^\alpha\| \le \eta$ for all $m \in U \cap M$, for $\alpha = c, u$, and s.

Remark. Because M is compact, for any $\eta > 0$ there exists $r > 0$ independent of $m \in M$ such that $B(m, r)$ is an η-neighborhood.

With these projections we can define local trivializations of the bundles:
For $m_0 \in M$, U a neighborhood of m_0, and for $\alpha = c, u$, and s, define

$$\Phi_{m_0}^\alpha : (U \cap M) \times X_{m_0}^\alpha \to X^\alpha$$

by

$$\Phi_{m_0}^\alpha(m, x^\alpha) = (m, \Pi_m^\alpha x^\alpha)$$

and

$$\Phi_{m_0}^{us} : (U \cap M) \times \left(X_{m_0}^u \oplus X_{m_0}^s \right) \to X^u \oplus X^s$$

by

$$\Phi_{m_0}^{us}(m, x^u + x^s) = (m, \Pi_m^u x^u + \Pi_m^s x^s).$$

In Section 4 we show that for $\epsilon > 0$ and sufficiently small Θ is a diffeomorphism from $X^u(\epsilon) \oplus X^s(\epsilon)$ onto a tubular neighborhood, V, of M. As such, we may define $\Psi \equiv \left(\Theta \big|_{X^u(\epsilon) \oplus X^s(\epsilon)} \right)^{-1} : V \to X^u(\epsilon) \oplus X^s(\epsilon)$ with component functions denoted through

$$\Psi(x) = (\psi(x), \psi^u(x) + \psi^s(x))$$

and

$$\psi^\alpha(x) = \Pi_{\psi(x)}^\alpha (x - \psi(x)).$$

In section 5 we prove that certain cone-like sets have invariance properties as a result of the normal hyperbolicity of M. These geometrical properties of the flow will then be used to establish the main results. Here we define and illustrate the various "cones" to be used.

For $\epsilon > \delta > 0$ and small and $\mu > 0$ we define:
The global cone

$$K(\epsilon, \mu) = \left\{ m + x^u + x^s : (m, x^u + x^s) \in \overline{X}^u(\epsilon) \oplus \overline{X}^s(\epsilon) \text{ and } \mu|x^u| \geq |x^s| \right\}$$

the fattened cone

$$K_f(\epsilon, \mu, \delta) = \{ m + x^u + x^s : (m, x^u + x^s) \in \overline{X}^u(\epsilon) \oplus \overline{X}^s(\epsilon) \text{ and }$$
$$\mu|x^u| \geq |x^s| \text{ or } |x^s| \leq \delta, \}$$

and the gap-cone

$$K_g(\epsilon, \mu, \delta) = \{ m + x^u + x^s : (m, x^u + x^s) \in \overline{X}^u(\epsilon) \oplus \overline{X}^s(\epsilon), \mu|x^u| \geq |x^s| \text{ and }$$
$$|x^u| \geq \delta \}.$$

The fattened cone and the gap cone are introduced to allow for perturbations.
The closures of their complements in the ϵ-neighborhood of M will be denoted with a prime. For instance,

$$K_f'(\epsilon, \mu, \delta) = \{ m + x^u + x^s : (m, x^u + x^s) \in \overline{X}^u(\epsilon) \oplus \overline{X}^s(\epsilon), \mu|x^u| \leq |x^s| \text{ and }$$
$$|x^s| \geq \delta \}.$$

Some of these cones are illustrated below.

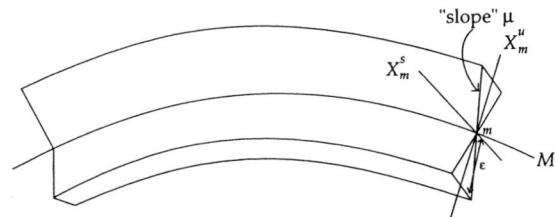

Fig 2 a) Global cone

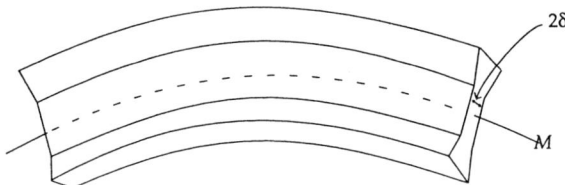

Fig 2 b) Fattened cone

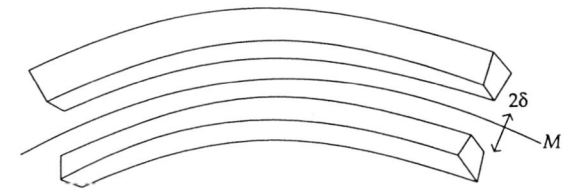

Fig 2 c) Gap cone

Figure 2

3. Statements of Theorems.

The principal result of this paper is the persistence of compact, smooth, normally hyperbolic invariant manifolds under C^1 perturbations of the semiflow. We shall, however, prove more than this by first considering *maps* which are C^1 perturbations, $\tilde{T}$, of the time-t map associated with the original semiflow. Secondly, we obtain, $\tilde{M}$, the perturbed compact invariant manifold by showing the existence of *center-unstable* and *center-stable* manifolds for $\tilde{T}$, in a neighborhood of M, and then taking their intersection. Furthermore, we give characterizations of the center-unstable and center-stable manifolds both for maps and perturbations of the semiflow.

Let B be a fixed neighborhood of M in X containing the tubular neighborhood $\Theta(X^u(\epsilon_0)) \oplus X^s(\epsilon_0))$. We consider a C^1 perturbation $\tilde{T}$ of the time-t map T^t. Let $\tilde{T} : B \to X$ be a C^1 map. Recall that

$$\|\tilde{T} - T^t\|_0 = \sup_{x \in B} |\tilde{T}(x) - T^t(x)|$$

and

$$\|\tilde{T} - T^t\|_1 = \sup_{x \in B} |\tilde{T}(x) - T^t(x)| + \sup_{x \in B} \|D\tilde{T}(x) - DT^t(x)\|.$$

Theorem A. *Let $t > t_0$ be fixed. For each small $\epsilon > 0$, there exists $\sigma > 0$ such that if $\|\tilde{T} - T^t\|_1 \leq \sigma$, then $\tilde{T}$ has a unique C^1 manifold $\tilde{W}^{cu}(\epsilon)$ in the tubular neighborhood, $\Theta(X^u(\epsilon) \oplus X^s(\epsilon))$ which satisfies*

(i) $\tilde{T}(\tilde{W}^{cu}(\epsilon)) \cap \Theta(X^u(\epsilon) \oplus X^s(\epsilon)) = \tilde{W}^{cu}(\epsilon)$
(ii) $\tilde{T} : \tilde{W}^{cu}(\epsilon) \cap \tilde{T}^{-1}(\tilde{W}^{cu}(\epsilon)) \to \tilde{W}^{cu}(\epsilon)$ *is a diffeomorphism,*
(iii) $\tilde{W}^{cu}(\epsilon) = \cap_{k=1}^{\infty} \mathcal{A}_k$, *where $\mathcal{A}_k$ is defined by induction, $\mathcal{A}_k = \tilde{T}(\mathcal{A}_{k-1}) \cap \Theta(X^u(\epsilon) \oplus X^s(\epsilon))$ and $\mathcal{A}_0 = \Theta(X^u(\epsilon) \oplus X^s(\epsilon))$.*

From property (iii) we see that the local center-unstable manifold, $\tilde{W}^{cu}(\epsilon)$, consists of points for which backward orbits exist and stay in the tubular neighborhood for all backward iterates. Likewise, the local center-stable manifold consists of points for which all forward iterates lie in the tubular neighborhood, as described in (ii) below. These, then, can be taken as definitions of the center-unstable and center-stable manifolds.

Theorem B. *Let $t > t_0$ be fixed. For each small $\epsilon > 0$, there exists $\sigma > 0$ such that if $\|\tilde{T} - T^t\|_1 \leq \sigma$, then $\tilde{T}$ has a unique C^1 manifold $\tilde{W}^{cs}(\epsilon)$ in the tubular neighborhood, $\Theta(X^u(\epsilon) \oplus X^s(\epsilon))$ which satisfies*

(i) $\tilde{T}(\tilde{W}^{cs}(\epsilon)) \subset \tilde{W}^{cs}(\epsilon)$,
(ii) $\tilde{W}^{cs}(\epsilon) = \cap_{k=1}^{\infty} \mathcal{A}_{-k}$, *where $\mathcal{A}_{-k}$ is defined by induction, $\mathcal{A}_{-k} = \left\{ x : \tilde{T}(x) \in \mathcal{A}_{1-k} \right\} \cap \Theta(X^u(\epsilon) \oplus X^s(\epsilon))$ and $\mathcal{A}_0 = \Theta(X^u(\epsilon) \oplus X^s(\epsilon))$.*

The above theorems combine to yield the existence of $\tilde{M}$, an invariant manifold which is a perturbation of M.

Theorem C. *Let $t > t_0$ be fixed. For each small $\epsilon > 0$, there exists $\sigma > 0$ such that if $\|\tilde{T} - T^t\|_1 \leq \sigma$, then $\tilde{T}$ has a unique C^1 compact connected normally hyperbolic invariant manifold $\tilde{M}$ in the tubular neighborhood, $\Theta(X^u(\epsilon) \oplus X^s(\epsilon))$ which satisfies*

    (i) $\tilde{M} = \tilde{W}^{cs}(\epsilon) \cap \tilde{W}^{cu}(\epsilon)$,

    (ii) $\tilde{T} : \tilde{M} \to \tilde{M}$ *is a C^1 diffeomorphism,*

    (iii) *There exists a C^1 diffeomorphism $K = K_{\tilde{T}} : M \to \tilde{M}$ which satisfies*

$$\|K_{\tilde{T}} - I\|_{C^1(M,X)} \to 0, \qquad as \quad \|\tilde{T} - T^t\|_1 \to 0.$$

    (iv) $\tilde{W}^{cs}(\epsilon)$ *and $\tilde{W}^{cu}(\epsilon)$ at $\tilde{M}$ are tangent to $\tilde{X}^s \oplus \tilde{X}^c$ and $\tilde{X}^u \oplus \tilde{X}^c$, respectively, where $X = \tilde{X}^u \oplus \tilde{X}^s \oplus \tilde{X}^c$ is the invariant decomposition associated with the normal hyperbolicity.*

We now may call $\tilde{W}^{cs}(\epsilon)$ and $\tilde{W}^{cu}(\epsilon)$ the (local) center-stable and center-unstable manifolds for $\tilde{T}$.

In addition to the characterizations of $\tilde{W}^{cu}(\epsilon)$ and $\tilde{W}^{cs}(\epsilon)$ given in Theorem A (iii) and Theorem B (ii), respectively, we actually have that $\tilde{W}^{cu}(\epsilon)$ is the unstable manifold of $\tilde{M}$ and $\tilde{W}^{cs}(\epsilon)$ is the stable manifold of $\tilde{M}$:

Theorem D.

$$\tilde{W}^{cu}(\epsilon) = \Big\{ x_0 \in \Theta(X^u(\epsilon) \oplus X^s(\epsilon)) : \text{ there exists } \{x_k\}_{k>0} \subset \Theta(X^u(\epsilon) \oplus X^s(\epsilon)),$$

$$such \; that \; T(x_k) = x_{k-1}, \; for \; k \geq 1, \; and \; x_k \to \tilde{M} \; as \; k \to \infty \Big\},$$

and

$$\tilde{W}^{cs}(\epsilon) = \Big\{ x_0 \in \Theta(X^u(\epsilon) \oplus X^s(\epsilon)) : \tilde{T}^k(x_0) \in \Theta(X^u(\epsilon) \oplus X^s(\epsilon)), k \geq 1,$$

$$and \; \tilde{T}^k(x_0) \to \tilde{M} \; as \; k \to \infty \Big\}.$$

We now consider the perturbed semiflow $\tilde{T}^t$ of T^t. Assume the perturbed semiflow $\tilde{T}^t$ is continuous on $[0, \infty) \times X$ into X, and for each $t \geq 0, \tilde{T}^t : X \to X$ is C^1.

For $t_1 > t_0$ we apply theorems A–C to the time-t_1 map $\tilde{T}^{t_1}$, to obtain $\tilde{W}^{cu}(\epsilon)$, $\tilde{W}^{cs}(\epsilon)$ and $\tilde{M}$ if $\|\tilde{T}^{t_1} - T^{t_1}\|_1$ is sufficiently small. The following results indicate that they are the center-stable, center-unstable, and normally hyperbolic invariant manifolds for the the semiflow $\tilde{T}^t$.

Theorem A'. *Let $t_1 > t_0$ be fixed. For each small $\epsilon > 0$, there exists $\sigma > 0$ such that if*

$$\|\tilde{T}^{t_1} - T^{t_1}\|_1 < \sigma \; and \; \|\tilde{T}^t - T^t\|_0 < \sigma, \; for \; 0 \leq t \leq t_1$$

then

    (i) $\tilde{T}^t(\tilde{W}^{cu}(\epsilon)) \cap \Theta(X^u(\epsilon) \oplus X^s(\epsilon)) \subset \tilde{W}^{cu}(\epsilon)$ *for $0 \leq t \leq t_1$,*

(ii) *For* $x \in \tilde{W}^{cu}(\epsilon)$, *if* $\tilde{T}^t(x) \in \Theta(X^u(\epsilon) \oplus X^s(\epsilon))$ *for* $0 \le t \le t_2$, *then* $\tilde{T}^t(x) \in \tilde{W}^{cu}(\epsilon)$ *for* $0 \le t \le t_2$,

(iii) $\tilde{W}^{cu}(\epsilon) \supset \cap_{t \ge 0} \tilde{\mathcal{A}}_t$, *where*

$$\tilde{\mathcal{A}}_t = \big\{ m_0 + x_0^u + x_0^s \in \Theta(X^u(\epsilon) \oplus X^s(\epsilon)) : \text{ there exists } (m_1, x_1^u + x_1^s) \in X^u(\epsilon) \oplus X^s(\epsilon)$$

$$\text{such that } \tilde{T}^t(m_1 + x_1^u + x_1^s) = m_0 + x_0^u + x_0^s,$$

$$\text{and } \tilde{T}^\tau(m_1 + x_1^u + x_1^s) \in \Theta(X^u(\epsilon) \oplus X^s(\epsilon)), \text{ for all } 0 \le \tau \le t \big\}.$$

Theorem B′. *Let* $t_1 > t_0$ *be fixed. For each small* $\epsilon > 0$, *there exists* $\sigma > 0$ *such that if*

$$\|\tilde{T}^{t_1} - T^{t_1}\|_1 < \sigma \text{ and } \|\tilde{T}^t - T^t\|_0 < \sigma, \text{ for } 0 \le t \le t_1$$

then

(i) *For all* $t \ge 0$, $\tilde{T}^t(\tilde{W}^{cs}(\epsilon)) \cap \Theta(X^u(\epsilon) \oplus X^s(\epsilon)) \subset \tilde{W}^{cs}(\epsilon)$,

(ii) $\big\{ m_0 + x_0^u + x_0^s : \tilde{T}^t(m_0 + x_0^u + x_0^s) \in \Theta(X^u(\epsilon) \oplus X^s(\epsilon)), t \ge 0 \big\} \subset \tilde{W}^{cs}(\epsilon)$.

Theorem C′. *Let* $t_1 > t_0$ *be fixed. For each small* $\epsilon > 0$, *there exists* $\sigma > 0$ *such that if*

$$\|\tilde{T}^{t_1} - T^{t_1}\|_1 < \sigma \text{ and } \|\tilde{T}^t - T^t\|_0 < \sigma, \text{ for } 0 \le t \le t_1$$

then $\tilde{M}$ *is a* C^1 *compact connected normally hyperbolic invariant manifold for* $\tilde{M}$ *and for each* $t \ge 0$, $\tilde{T}^t$ *is a* C^1 *diffeomorphism from* $\tilde{M}$ *onto* $\tilde{M}$.

Theorem D′.

(i) *For each* $x \in \tilde{W}^{cu}(\epsilon)$, $\lim_{t \to \infty} d(\tilde{T}^{-t}(x), \tilde{M}) = 0$, *uniformly on* $\tilde{W}^{cu}(\epsilon)$
    and

(ii) *For each* $x \in \tilde{W}^{cs}(\epsilon)$, $\lim_{t \to \infty} d(\tilde{T}^t(x), \tilde{M}) = 0$, *uniformly on* $\tilde{W}^{cs}(\epsilon)$

Remark. In trying to apply these theorems it is sometimes helpful to know how σ depends upon ϵ. In the case that T^t is C^2 for $t \ge 0$, then one can show that σ can be taken to be of order ϵ^s for any $s > 1$.

4. Local Coordinate Systems.

In this section, we first prove that the bundles introduced in Section 2 are C^1 and that we can find a tubular neighborhood of M which is C^1–diffeomorphic to open set in $X^u \oplus X^s$. Then we introduce three coordinate systems. The first coordinate system is given by the tubular neighborhood. The second is defined by the splitting of the tangent bundle of the phase space X restricted to M into its center, stable, and unstable components, as mentioned in (H3), which is a Cartesian coordinate system in a neighborhood of a point on M. The third coordinate system is induced by the local trivialization of the bundles based on M. Then we estimate how the coordinates differ from each other.

Let us first establish the relations among corresponding fibers X_m^α, $\alpha = u, s$, and c, as the base point moves through M.

Lemma 4.1. *For $m_1, m_2 \in M$, $X_{m_1}^\alpha$ is isomorphic to $X_{m_2}^\alpha$ for $\alpha = u, s, c$.*

Proof. From the hypothesis (H4), Π_m^α is continuous in m, thus for fixed m and $0 < \eta < 1$ there exists an η-neighborhood V_m of m in X, that is,

$$\|\Pi_{\bar{m}}^\alpha - \Pi_m^\alpha\| \leq \eta, \ \alpha = u, s, c \tag{4.1}$$

for any $\bar{m} \in V_m \cap M$.

We first claim that if $\eta < \sqrt{2} - 1$, then X_m^α is isomorphic to $X_{\bar{m}}^\alpha$ for $\bar{m} \in V_m \cap M$. To see this, let $x \in X_m^\alpha$, then (4.1) implies

$$|\Pi_{\bar{m}}^\alpha x - x| = |\Pi_{\bar{m}}^\alpha x - \Pi_m^\alpha x| \leq \eta|x|, \tag{4.2}$$

which yields

$$|\Pi_{\bar{m}}^\alpha x| \leq (1 + \eta)|x|. \tag{4.3}$$

Similarly, for $y \in X_{\bar{m}}^\alpha$

$$|\Pi_m^\alpha y - y| = |\Pi_m^\alpha y - \Pi_{\bar{m}}^\alpha y| \leq \eta|y|.$$

Thus, for $x \in X_m^\alpha$

$$\begin{aligned}
&|\Pi_m^\alpha \Pi_{\bar{m}}^\alpha x - x| \\
&\leq |\Pi_m^\alpha \Pi_{\bar{m}}^\alpha x - \Pi_{\bar{m}}^\alpha x| + |\Pi_{\bar{m}}^\alpha x - x| \\
&\leq \eta|\Pi_{\bar{m}}^\alpha x| + \eta|x| \\
&\leq \eta(2 + \eta)|x|,
\end{aligned}$$

which implies

$$\|I - \Pi_m^\alpha \Pi_{\bar{m}}^\alpha |_{X_m^\alpha}\| \leq \eta(2 + \eta) .$$

It follows that, as long as $\eta < \sqrt{2} - 1$, $\Pi_m^\alpha \Pi_{\bar{m}}^\alpha \big|_{X_m^\alpha}$ is invertible. Hence $\Pi_m^\alpha \Pi_{\bar{m}}^\alpha \big|_{X_m^\alpha}$ is an isomorphism. Interchanging m and $\bar{m}$, we obtain that $\Pi_{\bar{m}}^\alpha \Pi_m \big|_{X_{\bar{m}}^\alpha}$ is an isomorphism. Therefore $\Pi_m^\alpha \big|_{X_{\bar{m}}^\alpha}$ and $\Pi_{\bar{m}}^\alpha \big|_{X_m^\alpha}$ are isomorphisms.

Note that (4.2) also yields for $x \in X_m^\alpha$

$$|\Pi_{\bar{m}}^\alpha x| \geq (1 - \eta)|x|. \tag{4.4}$$

Therefore we have the estimates

$$\left\| \Pi_{\bar{m}}^\alpha \big|_{X_m^\alpha} \right\| \leq 1 + \eta \ \text{ and } \ \left\| \left(\Pi_{\bar{m}}^\alpha \big|_{X_m^\alpha} \right)^{-1} \right\| \leq \frac{1}{1 - \eta} \tag{4.5}$$

Since M is compact, we conclude that for any $m_1, m_2 \in M$, $X_{m_1}^\alpha$ is isomorphic to $X_{m_2}^\alpha$. This completes the proof. $\qquad\square$

As we mentioned in Section 2, for $(m, x^\alpha) \in X^\alpha$, $\alpha = u, s, c$, we may define a "norm" for the bundle X^α for $\alpha = u, s, c$, by

$$\|(m, x^\alpha)\| = |x^\alpha|.$$

Under this norm, we have

Lemma 4.2. X^α for $\alpha = u, s, c$, is a C^1 bundle with the local trivialization $\Phi_{m_0}^\alpha$ (defined in Section 2).

Proof. For each $m_1, m_2 \in M$, let V_{m_1} and V_{m_2} be η−neighborhoods of m_1 and of m_2, respectively, with $\eta < \sqrt{2} - 1$. Let $V = V_{m_1} \cap V_{m_2} \cap M$ and suppose $V \neq \phi$. Consider

$$\left(\Phi_{m_2}^\alpha \right)^{-1} \Phi_{m_1}^\alpha : V \times X_{m_1}^\alpha \to V \times X_{m_2}^\alpha.$$

We first notice that from (4.5), $\left(\Phi_{m_2}^\alpha \right)^{-1} \Phi_{m_1}^\alpha$ is well-defined.

Furthermore, $\left(\Phi_{m_2}^\alpha \right)^{-1} \Phi_{m_1}^\alpha (m, \cdot)$ is an isomorphism from $X_{m_1}^\alpha$ to $X_{m_2}^\alpha$ and the image of (m, x_1^α) is given by

$$(m, x_2^\alpha) = \Phi_{m_2}^{-1} \Phi_{m_1} (m, x_1^\alpha),$$

where

$$x_2^\alpha = x_2^\alpha(m) = \left(\Pi_m^\alpha \big|_{X_{m_2}^\alpha} \right)^{-1} \left(\Pi_m^\alpha \big|_{X_{m_1}^\alpha} \right) x_1^\alpha. \tag{4.6}$$

Let us first formally compute the derivative of $x_2^\alpha(m)$ in m. Applying the projection Π_m^α to (4.6), we obtain

$$\Pi_m^\alpha x_2^\alpha(m) = \Pi_m^\alpha x_1^\alpha. \tag{4.7}$$

Formally computing the derivative of (4.7) in m, we have

$$\Pi_m^\alpha D x_2^\alpha(m) + (D\Pi_m^\alpha) x_2^\alpha(m) = D\Pi_m^\alpha x_1^\alpha.$$

Thus

$$Dx_2^\alpha(m) = \left(\Pi_m^\alpha \Big|_{X_{m_2}^\alpha}\right)^{-1} (D\Pi_m^\alpha) (x_1^\alpha - x_2^\alpha(m)). \tag{4.8}$$

It is clear from (4.5) that the right hand side of (4.8) is well-defined.

Next we shall show that the right hand side of (4.8) indeed is the derivative of $x_2^\alpha(m)$.

Observe that $x_2^\alpha(m)$ is C^0 in m uniformly with respect to unit vectors x_1^α. In fact, from (4.7) we have

$$\Pi_{\bar{m}}^\alpha (x_2^\alpha(\bar{m}) - x_2^\alpha(m)) = (\Pi_{\bar{m}}^\alpha - \Pi_m^\alpha) (x_1^\alpha - x_2^\alpha(m)).$$

Thus, from the continuity of Π_m^α in m and the fact that $\Pi_{\bar{m}}^\alpha|_{X_{m_2}^\alpha}$ is invertible, it follows that $x_2^\alpha(m)$ is C^0 in m uniformly with respect to unit vectors x_1^α.

In order to prove $x_2^\alpha(m)$ is C^1, since M is a C^2 manifold, we choose a C^2 local chart $(U_m \cap M, \phi)$, where

$$\phi : B(0,1) \subset \mathbb{R}^n \to U_m \cap M$$

and $\phi(0) = m$.

We first note that

$$\begin{aligned}
&\Pi_{\phi(0)}^\alpha (x_2^\alpha(\phi(\tau v^*)) - x_2^\alpha(\phi(0))) \\
&= \Pi_{\phi(\tau v^*)}^\alpha (x_2^\alpha(\phi(\tau v^*)) - x_2^\alpha(\phi(0))) \\
&\quad - \left(\Pi_{\phi(\tau v^*)}^\alpha - \Pi_{\phi(0)}^\alpha\right) (x_2^\alpha(\phi(\tau v^*)) - x_2^\alpha(\phi(0))).
\end{aligned}$$

Thus, by using (4.7), we find the identity

$$\Pi_{\phi(\tau v^*)}^\alpha(x_2^\alpha(\phi(\tau v^*)) - x_2^\alpha(\phi(0))) = -(\Pi_{\phi(\tau v^*)}^\alpha - \Pi_{\phi(0)}^\alpha)(x_2^\alpha(\phi(0)) - x_1^\alpha).$$

Since Π_m^α is C^1 in m by hypotheses (H4), using the Taylor expansion, we obtain

$$\begin{aligned}
&\left(\Pi_{\phi(\tau v^*)}^\alpha - \Pi_{\phi(0)}^\alpha\right) (x_2^\alpha(\phi(0)) - x_1^\alpha) \\
&= \tau D\Pi_{\phi(0)}^\alpha(D\phi \, v^*)(x_2^\alpha(\phi(0)) - x_1^\alpha) + \tau |x_2^\alpha(\phi(0)) - x_1^\alpha| O(\tau),
\end{aligned}$$

where the quantity $O(\tau)$ does not depend on v^* or x_1^α and $O(\tau) \to 0$ as $\tau \to 0$. Hence,

$$\begin{aligned}
&\Pi_{\phi(0)}^\alpha \frac{1}{\tau} (x_2^\alpha(\phi(\tau v^*)) - x_2^\alpha(\phi(0))) \\
&= -D\Pi_{\phi(0)}^\alpha(D\phi \, v^*) (x_2^\alpha(\phi(0)) - x_1^\alpha) - |x_2^\alpha(\phi(0)) - x_1^\alpha| O(\tau) \\
&\quad - \frac{1}{\tau} \left(\Pi_{\phi(\tau v^*)}^\alpha - \Pi_{\phi(0)}^\alpha\right) (x_2^\alpha(\phi(\tau v^*)) - x_2^\alpha(\phi(0))). \tag{4.9}
\end{aligned}$$

Since $x_2^\alpha(\phi(\tau v^*)) \to x_2^\alpha(\phi(0))$ as $\tau \to 0$ uniformly in v^* and x_1^α, and Π_m^α is C^1 in m, letting $\tau \to 0$ in (4.9), we find

$$\Pi_{\phi(0)}^\alpha \frac{1}{\tau} \left(x_2^\alpha(\phi(\tau v^*)) - x_2^\alpha(\phi(0)) \right) \to -D\Pi_{\phi(0)}^\alpha D\phi \, v^* \left(x_2^\alpha(\phi(0)) - x_1^\alpha \right)$$

uniformly in v^* and x_1^α. Therefore, $x_2^\alpha(m)$ is differentiable and (4.8) holds since $\Pi_{\phi(0)}^\alpha |_{X_{m_2}^\alpha}$ is invertible.

Finally, we show that $Dx_2^\alpha(m)$ is continuous in m. We first notice that (4.8) may also be written as

$$Dx_2^\alpha(m) = \left(\Pi_m^\alpha \Big|_{X_{m_2}^\alpha} \right)^{-1} \Pi_m^\alpha (D\Pi_m^\alpha) \left(x_1^\alpha - x_2^\alpha(m) \right).$$

Since $D\Pi_m^\alpha$ is continuous in m from (H4) and $(x_1^\alpha - x_2^\alpha(m))$ is continuous in m uniformly with respect to unit vectors x_1^α, it is sufficient to prove that

$$\left(\Pi_m^\alpha \Big|_{X_{m_2}^\alpha} \right)^{-1} \Pi_m^\alpha$$

is continuous in m.

Let $x \in X$ and set

$$y = \left(\Pi_m^\alpha \Big|_{X_{m_2}^\alpha} \right)^{-1} \Pi_m^\alpha x \quad \text{and} \quad \bar{y} = \left(\Pi_{\bar{m}}^\alpha \Big|_{X_{m_2}^\alpha} \right)^{-1} \Pi_{\bar{m}}^\alpha x.$$

We obtain

$$|\Pi_m^\alpha x - \Pi_{\bar{m}}^\alpha x|$$
$$= |\Pi_m^\alpha y - \Pi_{\bar{m}}^\alpha \bar{y}|$$
$$\geq |\Pi_{\bar{m}}^\alpha (y - \bar{y})| - \|\Pi_m^\alpha - \Pi_{\bar{m}}^\alpha\||y|.$$

Thus,

$$|\Pi_{\bar{m}}^\alpha (y - \bar{y})| \leq \|\Pi_m^\alpha - \Pi_{\bar{m}}^\alpha\|(|x| + |y|) \tag{4.10}$$

On the other hand, from (4.5) it follows that

$$|y - \bar{y}| = |\left(\Pi_{\bar{m}}^\alpha \Big|_{X_{m_2}^\alpha} \right)^{-1} \Pi_{\bar{m}}^\alpha (y - \bar{y})|$$
$$\leq \|\left(\Pi_{\bar{m}}^\alpha \Big|_{X_{m_2}^\alpha} \right)^{-1}\||\Pi_{\bar{m}}^\alpha (y - \bar{y})|$$
$$\leq \frac{1}{1-\eta}|\Pi_{\bar{m}}^\alpha (y - \bar{y})|.$$

We also note that

$$|y| \leq \frac{C}{1-\eta}|x|,$$

where C is a bound for $\|\Pi_m^\alpha\|$. Therefore,

$$|y - \bar{y}| \le \frac{1 + C - \eta}{(1 - \eta)^2}\|\Pi_m^\alpha - \Pi_{\bar{m}}^\alpha\||x|$$

which implies

$$\left\|\left(\Pi_m^\alpha\Big|_{X_{m_2}^\alpha}\right)^{-1}\Pi_m^\alpha - \left(\Pi_{\bar{m}}^\alpha\Big|_{X_{m_2}^\alpha}\right)^{-1}\Pi_{\bar{m}}^\alpha\right\| \le \frac{1 + C - \eta}{(1 - \eta)^2}\|\Pi_m^\alpha - \Pi_{\bar{m}}^\alpha\|. \qquad (4.11)$$

This gives the continuity of $\left(\Pi_m^\alpha\Big|_{X_{m_2}^\alpha}\right)^{-1}\Pi_m^\alpha$ in m and completes the proof. $\qquad \square$

From Lemma 4.2 one easily sees that $X^u \oplus X^s$ is a C^1 bundle.

The next result implies that the image of $X^u(\epsilon) \oplus X^s(\epsilon)$ is a tubular neighborhood of M for ϵ sufficiently small.

Lemma 4.3. *There exists $\epsilon > 0$ such that Θ is a C^1 diffeomorphism from $X^u(\epsilon) \oplus X^s(\epsilon)$ onto a neighborhood of M.*

Proof. For $m_0 \in M$, let $B(m_0, r)$ be an η-neighborhood of m_0 with $\eta < \sqrt{2} - 1$. Consider the map

$$\Theta\Phi_{m_0}^{us} : B(m_0, r) \cap M \times \left(X_{m_0}^u \oplus X_{m_0}^s\right) \to X.$$

Let $(x_1^c, x_1^u + x_1^s)$ be a tangent vector at $(m, 0)$ to the manifold $B(m_0, r) \cap M \times \left(X_{m_0}^u \oplus X_{m_0}^s\right)$. A simple computation gives

$$D(\Theta\Phi_{m_0}^{us})\big|_{(m,0)}(x^c, x_1^u + x_1^s) = x^c + \Pi_m^u x_1^u + \Pi_m^s x_1^s.$$

Since $\Pi_m^u\big|_{X_{m_0}^u}$ and $\Pi_m^s\big|_{X_{m_0}^s}$ are isomorphisms from (4.5), $D(\Theta\Phi_{m_0}^{us})$ is an isomorphism from $X_{m_0}^c \times \left(X_{m_0}^u \oplus X_{m_0}^s\right)$ to X. Therefore, by the Inverse Function Theorem, $\Theta\Phi_{m_0}^{us}$ is a diffeomorphism from a neighborhood of $(m, 0) \in B(m_0, r) \cap M \times \left(X_{m_0}^u \oplus X_{m_0}^s\right)$ onto a neighborhood of m in X. Hence, Θ is a diffeomorphism from a neighborhood of $(m, 0) \in X^u \oplus X^s$ onto a neighborhood of m in X.

Because of the compactness of M, there is a finite covering of M,

$$B(m_i, \epsilon_i) = \{m \in M : |m - m_i| < \epsilon_i\}, \ i = 1, 2, \cdots, k,$$

where ϵ_i are positive constants chosen such that Θ restricted on each subset

$$E(m_i, \epsilon_i) = \{(m, x^u + x^s) \in X^u(\epsilon_i) \oplus X^s(\epsilon_i) : m \in B(m_i, 5\epsilon_i)\}$$

is a diffeomorphism. Let $\epsilon = \min\{\epsilon_i\}$. To show that Θ is a diffeomorphism from $X^u(\epsilon) \times X^s(\epsilon)$ onto a neighborhood of M in X, it is sufficient to prove that Θ is one-to-one. Let $(m, x^u + x^s)$, $(\bar{m}, \bar{x}^u + \bar{x}^s) \in X^u(\epsilon) \oplus X^s(\epsilon)$. We first notice that $m \in B(m_i, \epsilon_i)$ for some i. Suppose

$$\Theta(m, x^u + x^s) = \Theta(\bar{m}, \bar{x}^u + \bar{x}^s),$$

i.e.,

$$m + x^u + x^s = \bar{m} + \bar{x}^u + \bar{x}^s.$$

Then

$$|m - \bar{m}| = |\bar{x}^u - x^u + \bar{x}^s - x^s|$$
$$< 4\epsilon,$$

which gives $\bar{m} \in B(m_i, 5\epsilon_i)$ and

$$(m, x^u + x^s), (\bar{m}, \bar{x}^u + \bar{x}^s) \in E(m_i, \epsilon_i),$$

hence $m = \bar{m}$, $x^u = \bar{x}^u$, $x^s = \bar{x}^s$. This completes the proof. □

Obviously, Lemma 4.3 holds for any $0 < \epsilon_1 < \epsilon$.

The next result concerns about the projection Π_m^c.

Lemma 4.4. *For any $\epsilon > 0$, there exists $\beta^* > 0$ such that if $m_0, m_1, m_2 \in M$ satisfy*

$$|m_1 - m_0| < \beta^* \ and \ |m_2 - m_0| < \beta^*$$

then

$$\frac{|m_1 - m_2 - \Pi_{m_0}^c(m_1 - m_2)|}{|m_1 - m_2|} < \epsilon. \tag{4.12}$$

Proof. For $m_0 \in M$, let $\phi_{m_0} : B(0, 1) \subset \mathbb{R}^n \to U_{m_0} \cap M$ be the C^2 coordinate map with $\phi_{m_0}(0) = m_0$. Choose $\beta_0^* > 0$ such that if $|m_1 - m_0| < \beta_0^*$ and $|m_2 - m_0| < \beta_0^*$ then $m_1, m_2 \in U_{m_0} \cap M$. Let

$$v_i = \phi_{m_0}^{-1}(m_i), \ i = 1, 2.$$

We find

$$m_1 - m_2 = \int_0^1 D\phi_{m_0}\left(\tau v_1 + (1 - \tau)v_2\right)(v_1 - v_2)d\tau$$
$$= o(1)(v_1 - v_2) + D\phi_{m_0}(0)(v_1 - v_2) \tag{4.13}$$

as $v_1, v_2 \to 0$. Applying the projection $\Pi_{m_0}^c$ to (4.13), we obtain

$$\Pi_{m_0}^c(m_1 - m_2) = D\phi_{m_0}(0)(v_1 - v_2) + \Pi_{m_0}^c(o(1)(v_1 - v_2)).$$

Subtracting the above identity from (4.13), we have

$$|m_1 - m_2 - \Pi^c_{m_0}(m_1 - m_2)|$$
$$= o(1)|v_1 - v_2|$$
$$= o(1)|m_1 - m_2|.$$

as $m_1, m_2 \to m_0$. From the fact that M is compact and Π^c_m is continuous in m, by a standard compactness argument, it follows that there exists $\beta^* > 0$ such that (4.12) holds. The proof is complete. $\qquad\square$

Let $\epsilon > 0$ be the same as in Lemma 4.3. For $m_0 \in M$, let $B(m_0, r)$ be an η-neighborhood of m_0, where r is chosen so that this is an η-neighborhood with

$$\eta < \sqrt{2} - 1 \text{ and } r < \beta^* \qquad (4.14)$$

and β^* is given by Lemma 4.4. The core of this section is to express each point in a tubular neighborhood in terms of three coordinate systems, which are introduced as follows. For each point in the tubular neighborhood $\Theta(X^u(\epsilon) \oplus X^s(\epsilon))$, Lemma 4.3 gives the first coordinate system $(m, x^u + x^s) \in X^u(\epsilon) \oplus X^s(\epsilon)$. The advantage of this coordinate systems is that it is a global coordinate in a neighborhood of $\mathcal{M}$, but the disadvantage is that the coordinate spaces vary with the base point m. For $m_0 \in \mathcal{M}$ such that $m \in B(m_0, r)$, since $X = X^u_{m_0} \oplus X^s_{m_0} \oplus X^c_{m_0}$ from (i) in hypothesis (H3), there is a unique $\bar{x}^u + \bar{x}^s + \bar{x}^c \in X^u_{m_0} \oplus X^s_{m_0} \oplus X^c_{m_0}$ such that

$$m + x^u + x^s = m_0 + \bar{x}^u + \bar{x}^s + \bar{x}^c,$$

giving a Cartesian coordinate system in a neighborhood of m_0. The advantage of this system is that the coordinate spaces do not depend on the base point m, which plays a crucial role in the existence and smoothness of invariant manifolds. We shall see this in the future sections. If we trivialize the bundle near m_0, we have the third coordinate system modified from the first one. From (4.5) we find there exists a unique $\tilde{x}^u + \tilde{x}^s \in X^u_{m_0} \oplus X^s_{m_0}$ such that

$$m + x^u + x^s = m + \Pi^u_m \tilde{x}^u + \Pi^s_m \tilde{x}^s.$$

This defines the third coordinate system $(m, \tilde{x}^u, \tilde{x}^s)$.

We now consider the relations among these coordinate systems. Take two points $(m_i, x^u_i + x^s_i) \in X^u_{m_i}(\epsilon) \oplus X^s_{m_i}(\epsilon)$ with $m_i \in B(m_0, r) \cap M$, $i = 1, 2$. Then we have the three representations

$$m_i + x^u_i + x^s_i$$
$$= m_0 + \bar{x}^u_i + \bar{x}^s_i + \bar{x}^c_i$$
$$= m_i + \Pi^u_{m_i} \tilde{x}^u_i + \Pi^s_{m_i} \tilde{x}^s_i, \qquad (4.15)$$

where $\bar{x}^\alpha_i \in X^\alpha_{m_0}$, $\alpha = u, s, c$, $\tilde{x}^\alpha_i \in X^\alpha_{m_0}$, $\alpha = u, s$.

The next result estimates how the coordinates differ from each other.

Lemma 4.5.

$$|\bar{x}_1^c - \bar{x}_2^c - (m_1 - m_2)|$$
$$\leq C\epsilon|m_1 - m_2| + C|m_i - m_0|\left(|\tilde{x}_1^u - \tilde{x}_2^u| + |\tilde{x}_1^s - \tilde{x}_2^s|\right) \tag{4.16}$$

and

$$|\bar{x}_1^\alpha - \bar{x}_2^\alpha - (\tilde{x}_1^\alpha - \tilde{x}_2^\alpha)|$$
$$\leq C\epsilon|m_1 - m_2| + C|m_i - m_0|\left(|\tilde{x}_1^u - \tilde{x}_2^u| + |\tilde{x}_1^s - \tilde{x}_2^s|\right) \tag{4.17}$$

for $i = 1, 2, \alpha = u, s$, where C is a positive constant which depends only on the projections.

In the remainder of this paper, we shall use C as a generic constant depending only on the projections and the time-t map T^t.

Proof. Using (4.15), we subtract the second point from the first point and obtain

$$\bar{x}_1^u - \bar{x}_2^u + \bar{x}_1^s - \bar{x}_2^s + \bar{x}_1^c - \bar{x}_2^c$$
$$= m_1 - m_2 + \Pi_{m_1}^u \tilde{x}_1^u - \Pi_{m_2}^u \tilde{x}_2^u + \Pi_{m_1}^s \tilde{x}_1^s - \Pi_{m_2}^u \tilde{x}_2^s. \tag{4.18}$$

Then, taking projection $\Pi_{m_0}^c$, we get

$$\bar{x}_1^c - \bar{x}_2^c = \Pi_{m_0}^c (m_1 - m_2) + (\Pi_{m_0}^c - \Pi_{m_1}^c)\Pi_{m_1}^u (\tilde{x}_1^u - \tilde{x}_2^u)$$
$$+ \Pi_{m_0}^c (\Pi_{m_1}^u - \Pi_{m_2}^u)\tilde{x}_2^u + (\Pi_{m_0}^c - \Pi_{m_1}^c)\Pi_{m_1}^s (\tilde{x}_1^s - \tilde{x}_2^s)$$
$$+ \Pi_{m_0}^c (\Pi_{m_1}^s - \Pi_{m_2}^s)\tilde{x}_2^s.$$

Applying Lemma 4.4, we obtain

$$|\bar{x}_1^c - \bar{x}_2^c - m_1 + m_2|$$
$$\leq \epsilon|m_1 - m_2| + \|\Pi_{m_0}^c - \Pi_{m_1}^c\|(\|\Pi_{m_1}^u\|\|\tilde{x}_1^u - \tilde{x}_2^u| + \|\Pi_{m_1}^s\|\|\tilde{x}_1^s - \tilde{x}_2^s|)$$
$$+ \|\Pi_{m_0}^c\|(\|\Pi_{m_1}^u - \Pi_{m_2}^u\| |\tilde{x}_2^u| + \|\Pi_{m_1}^s - \Pi_{m_2}^s\| |\tilde{x}_2^s|)$$
$$\leq (\epsilon + \|\Pi_{m_0}^c\|(\text{Lip}(\Pi_m^u)|\tilde{x}_2^u| + \text{Lip}(\Pi_m^s)|\tilde{x}_2^s|)) |m_1 - m_2|$$
$$+ \text{Lip}(\Pi_m^c)|m_0 - m_1|(\|\Pi_{m_1}^u\|\|\tilde{x}_1^u - \tilde{x}_2^u| + \|\Pi_{m_1}^s\|\|\tilde{x}_1^s - \tilde{x}_2^s|).$$

Here we use the fact that Π_m^α is globally Lipschitz continuous in m in the operator norm, which comes from the hypothesis (H4).

It follows from (4.5) that $|\tilde{x}_2^u| \leq \frac{1}{1-\eta}\epsilon$ and $|\tilde{x}_2^s| \leq \frac{1}{1-\eta}\epsilon$. W also notice that $\eta \leq 1/2$. Hence, by letting $Q = \max\{\|\Pi_m^u\|, \|\Pi_m^s\|, \|\Pi_m^c\| : m \in M\}$ and

$$C = \max\{1 + 2Q(\text{Lip}(\Pi_m^u) + \text{Lip}(\Pi_m^s)), Q\text{Lip}(\Pi_m^c)\}$$

gives the proof of (4.16) for $i = 1$. Interchanging indices 1 and 2 in (4.18) gives (4.16) for $i = 2$.

Similarly, applying the projection $\Pi^u_{m_0}$ to (4.18), we have

$$\begin{aligned}
\bar{x}^u_1 - \bar{x}^u_2 &= \Pi^u_{m_0}(m_1 - m_2 - \Pi^c_{m_0}(m_1 - m_2)) \\
&\quad + \Pi^u_{m_0}\Pi^u_{m_1}(\tilde{x}^u_1 - \tilde{x}^u_2) + \Pi^u_{m_0}(\Pi^u_{m_1} - \Pi^u_{m_2})\tilde{x}^u_2 \\
&\quad + (\Pi^u_{m_0} - \Pi^u_{m_1})\Pi^s_{m_1}(\tilde{x}^s_1 - \tilde{x}^s_2) + \Pi^u_{m_0}(\Pi^s_{m_1} - \Pi^s_{m_2})\tilde{x}^s_2.
\end{aligned}$$

Hence, using (4.5),

$$\begin{aligned}
&|\bar{x}^u_1 - \bar{x}^u_2 - (\tilde{x}^u_1 - \tilde{x}^u_2)| \\
&\leq \|\Pi^u_{m_0}\|(\epsilon + \mathrm{Lip}(\Pi^s_m)|\tilde{x}^s_2| + \mathrm{Lip}(\Pi^u_m)|\tilde{x}^u_2|)|m_1 - m_2| \\
&\quad + |m_1 - m_0|\mathrm{Lip}(\Pi^u_m)(\|\Pi^u_{m_0}\|\|\tilde{x}^u_1 - \tilde{x}^u_2\| + \|\Pi^s_{m_1}|x^s_{m_0}\|\|\tilde{x}^s_1 - \tilde{x}^s_2\|) \\
&\leq \epsilon Q\{\|\Pi^u_m\|\}(1 + 2(\mathrm{Lip}(\Pi^s_m) + \mathrm{Lip}(\Pi^s_m))|m_1 - m_2| \\
&\quad + Q\mathrm{Lip}(\Pi^u_m))|m_1 - m_0|(|\tilde{x}^u_1 - \tilde{x}^u_2| + |\tilde{x}^s_1 - \tilde{x}^s_2|).
\end{aligned}$$

Thus, setting

$$C = \max\{Q\{\|\Pi^u_m\|\}(1 + 2(\mathrm{Lip}(\Pi^s_m) + \mathrm{Lip}(\Pi^s_m))), Q\mathrm{Lip}(\Pi^u_m)\}$$

gives (4.17) for $i = 1$ and $\alpha = u$. Similarly, one has that (4.17) holds for $i = 1$, $\alpha = s$ and for $i = 2$ and $\alpha = u, c$. $\square$

5. Cone Lemmas.

In Section 2 we defined several cones. We show in this section that for fixed $t \geq t_0$ they are invariant under the time-t map T^t provided ϵ and δ are sufficiently small, due to normally hyperbolicity. Furthermore, the global fattened cone and the global gap cone are invariant for a small C^1 perturbation $\tilde{T}$ of T^{t_1}. The invariance of local cones moving with the semiflow is also proved. These invariances play crucial roles for establishing the existence of center-unstable and center-stable manifolds.

We shall confine our study to a tubular neighborhood of M. Let $\epsilon_0 > 0$ be fixed such that $\Theta(X^u(\epsilon_0) \oplus X^s(\epsilon_0))$ is a tubular neighborhood contained in the fixed neighborhood B of M introduced in Section 3. We shall also work with a smaller tubular neighborhood $\Theta(X^u(\epsilon) \oplus X^s(\epsilon))$ such that its image under the time-t map T^t stays in the fixed tubular neighborhood $\Theta(X^u(\epsilon_0) \oplus X^s(\epsilon_0))$, which is assured by the next lemma. In fact, we shall work in such a neighborhood with ϵ even smaller.

Lemma 5.1. *Let* $0 \leq t_1 < \infty$. *For the fixed* ϵ_0, *there exists* $\epsilon^* > 0$ *such that for* $0 < \epsilon < \epsilon^*$

$$T^t\left(\Theta(\overline{X^u(\epsilon)} \oplus \overline{X^s(\epsilon)})\right) \subset \Theta(X^u(\epsilon_0) \oplus X^s(\epsilon_0)) \ for \ 0 \leq t \leq t_1. \tag{5.1}$$

The proof of this lemma, which we omit, is based on the continuity of T^t and the compactness of M.

Throughout the remainder of this paper we shall assume that ϵ satisfies (5.1)

The next lemma gives the invariance of the cones under the unperturbed time-t map T^t.

Lemma 5.2. *Let* $t \geq t_0$ *be fixed. For any* $\lambda_1 \in (\lambda, 1)$ *and* $\mu > 0$, *there exists* $\epsilon^* > 0$ *such that for each* $\epsilon \in (0, \epsilon^*)$ *and* $\delta \in (0, \epsilon)$, *the following statements hold:*

  (i) $T^t\left(K(\epsilon, \mu)\right) \cap \Theta(\overline{X^u(\epsilon)} \oplus \overline{X^s(\epsilon)}) \subset K(\epsilon, \lambda_1^2 \mu)$,

  (ii) $T^t\left(K_g(\epsilon, \mu, \delta)\right) \cap \Theta(\overline{X^u(\epsilon)} \oplus \overline{X^s(\epsilon)}) \subset K_g(\epsilon, \lambda_1^2 \mu, \lambda_1^{-1} \delta)$,

  (iii) $T^t\left(K_f(\epsilon, \mu, \delta)\right) \cap \Theta(\overline{X^u(\epsilon)} \oplus \overline{X^s(\epsilon)}) \subset K_f(\epsilon, \lambda_1^2 \mu, \lambda_1 \delta)$.

Proof. We first define a function

$$F : X \times X \to X$$

by

$$F(x, y) = \frac{T^t(y) - T^t(x) - DT^t(x)(y - x)}{|y - x|}, \quad \text{for } x \neq y,$$

and

$$F(x, x) = 0.$$

From the smoothness of T^t it follows that $F(x, y)$ is a continuous function. Let $(m_0, x_0^u + x_0^s) \in \overline{X^u(\epsilon)} \oplus \overline{X^s(\epsilon)}$. By Lemma 5.1, we have that for any fixed $\epsilon_2 < \epsilon_0$ as long as ϵ is sufficiently small, then

$$T^t\left(\Theta(\overline{X^u(\epsilon)} \oplus \overline{X^s(\epsilon)})\right) \subset \Theta(X^u(\epsilon_2) \oplus X^s(\epsilon_2)).$$

we may write

$$T^t(m_0 + x_0^u + x_0^s) = m_1 + x_1^u + x_1^s,$$

where $x_1^\alpha \in X_{m_1}^\alpha(\epsilon_2)$. By the Taylor expansion, we find

$$m_1 + x_1^u + x_1^s = T^t(m_0) + DT^t(m_0)(x_0^u + x_0^s) + F(m_0, x_0^u + x_0^s + m_0)|x_0^u + x_0^s|. \quad (5.2)$$

Therefore,

$$|m_1 - T^t(m_0)|$$
$$\leq |x_1^u| + |x_1^s| + \|DT^t(m_0)\|(|x_0^u| + |x_0^s|)$$
$$+ |F(m_0, x_0^u + x_0^s + m_0)|(|x_0^u + x_0^s|)$$
$$\leq 2\epsilon_2 + 2\epsilon C + 2\epsilon|F(m_0, x_0^u + x_0^s + m_0)|.$$

The fact that F is continuous in a neighborhood of the compact manifold $M \times M$ implies that $|F(m_0, x_0^u + x_0^s + m_0)| = O(\epsilon)$ uniformly with respect to $m_0,, x_0^u$, and x_0^s.

In order to apply the estimates of Lemma 4.5, we first need to show that m_1 is in an η-neighborhood of $T^t(m_0)$. By Lemma 4.4, for $\epsilon_1 < \epsilon_0$ there exists $\beta^* > 0$ such that (4.12) holds with ϵ_1 instead of ϵ. Let η, r satisfy (4.14).

Let $\epsilon_1 < \frac{1}{2C}$ where C is the constant in Lemma 4.5. Choose ϵ_2 such that $\epsilon_2 \leq \frac{1}{4}r$ and choose ϵ^* to satisfy the requirement that Lemma 5.1 holds and for any $\epsilon \in (0, \epsilon^*)$

$$2\epsilon C + 2\epsilon|F(m_0, x_0^u + x_0^s + m_0)| \leq \frac{r}{2}. \quad (5.3)$$

Thus, we obtain

$$|m_1 - T^t(m_0)| \leq r, \quad (5.4)$$

which implies that m_1 is in an η-neighborhood of $\bar{m} = T^t(m_0)$. Hence, we may write $m_1 + x_1^u + x_1^s$ as

$$m_1 + x_1^u + x_1^s = m_1 + \Pi_{m_1}^u \tilde{x}_1^u + \Pi_{m_1}^s \tilde{x}_1^s, \quad (5.5)$$

where $\tilde{x}_1^\alpha \in X_{\bar{m}}^\alpha$ for $\alpha = u, s$.

Applying Lemma 4.5 to points $\bar{m}$ and $m_1 + x_1^u + x_1^s$ and using (5.2) and (5.4), we obtain

$$|\Pi_{\bar{m}}^c F(m_0, x_0^u + x_0^s + m_0)|x_0^u + x_0^s| - (m_1 - \bar{m})| \leq C\epsilon_1|m_1 - \bar{m}|$$

which, since $\epsilon_1 < \frac{1}{2C}$, yields

$$|m_1 - \bar{m}| \leq 2|\Pi_{\bar{m}}^c F(m_0, x_0^u + x_0^s + m_0)||x_0^u + x_0^s|$$
$$= O(\epsilon)|x_0^u + x_0^s|$$

and

$$|DT^t(m)x_0^u + \Pi_{\bar{m}}^u F(m_0, x_0^u + x_0^s + m_0)|x_0^u + x_0^s| - \tilde{x}_1^u| \leq C\epsilon_1|m_1 - \bar{m}|.$$

This implies

$$|\tilde{x}_1^u - DT^t(m)x_0^u| \leq C\epsilon_1|m_1 - \bar{m}| + |\Pi_{\bar{m}}^u F(m_0, x_0^u + x_0^s + m_0)||x_0^u + x_0^s|$$
$$\leq C\epsilon_1|m_1 - \bar{m}| + O(\epsilon)|x_0^u + x_0^s|.$$

Thus,

$$\tilde{x}_1^u - DT^t(\bar{m})x_0^u = O(\epsilon)(|x_0^u| + |x_0^s|). \tag{5.6}$$

Applying the projection $\Pi_{m_1}^u$ to (5.6),

$$x_1^u = \Pi_{m_1}^u \tilde{x}_1^u = \Pi_{m_1}^u DT^t(m_0)x_0^u + O(\epsilon)\left(|x_0^u| + |x_0^s|\right). \tag{5.7}$$

Similarly,

$$x_1^s = \Pi_{m_1}^s \tilde{x}_1^s = \Pi_{m_1}^s DT^t(m_0)x_0^s + O(\epsilon)\left(|x_0^u| + |x_0^s|\right). \tag{5.8}$$

From (5.7) it follows that

$$|x_1^u| \geq \left|DT^t(m_0)|_{X_{m_0}^u} x_0^u\right| - \left|\left(\Pi_{m_1}^u - \Pi_{\bar{m}}^u\right) DT^t(m_0)x_0^u\right| - O(\epsilon)\left(|x_0^u| + |x_0^s|\right)$$
$$\geq \inf_{|\hat{x}_0^u|=1}\left|DT^t(m_0)|_{X_{m_0}^u}\hat{x}_0^u\right||x_0^u| - O(\epsilon)\left(|x_0^u| + |x_0^s|\right). \tag{5.9}$$

Similarly, from (5.8), we have

$$|x_1^s| \leq \left\|DT^t(m_0)\right|_{X_{m_0}^s}\||x_0^s| + O(\epsilon)\left(|x_0^u| + |x_0^s|\right). \tag{5.10}$$

For $(m_0, x_0^u + x_0^s) \in K(\epsilon, \mu)$, by using the normal hyperbolicity and (5.9), (5.10), we obtain

$$\lambda_1^2\mu|x_1^u| - |x_1^s|$$
$$\geq \lambda_1^2\mu \inf_{|\hat{x}_0^u|=1}|DT^t(m_0)|_{X_{m_0}^u}\hat{x}_0^u||x_0^u| - \|DT^t(m_0)|_{X_{m_0}^s}\||x_0^s| - O(\epsilon)\left(|x_0^u| + |x_0^s|\right)$$
$$\geq |x_0^s|\left(\left(\lambda_1^2 - \lambda^2\right) \inf_{|\hat{x}_0^u|=1}|DT^t(m_0)\hat{x}_0^u| - O(\epsilon)\right).$$

Since $\lambda_1 > \lambda$ and $\inf_{|\hat{x}_0^u|=1}|DT^t(m_0)\hat{x}_0^u| > \frac{1}{\lambda}$, we further require that ϵ^* is chosen to satisfy for $\epsilon < \epsilon^*$

$$\left(\lambda_1^2 - \lambda^2\right) \inf_{|\hat{x}_0^u|=1}|DT^t(m_0)\hat{x}_0^u| - O(\epsilon) > 0,$$

which implies that (i) holds.

To see that (ii) holds, it is enough to show that $|x_1^u| \geq \lambda_1^{-1}\delta$. From (5.9) and the normal hyperbolicity we have

$$|x_1^u| \geq \lambda^{-1}|x_0^u| - O(\epsilon)(1 + \mu)|x_0^u|$$
$$= \left(\lambda^{-1} - O(\epsilon)\right)|x_0^u|.$$

Thus by choosing smaller ϵ^* if necessary, we obtain $|x_1^u| \geq \lambda_1^{-1}\delta$

To see that (iii) holds, since (i) holds, it is enough to show that if $|x_0^s| \leq \delta$ and $\mu|x_1^u| < |x_1^s|$ then $|x_1^s| \leq \lambda_1\delta$. This directly follows from (5.10) and the normal hyperbolicity. In fact

$$|x_1^s| \leq \lambda|x_0^s| + O(\epsilon)\left(1 + \frac{1}{\mu}\right)|x_0^s|.$$

Thus, requireing ϵ^* to be smaller if necessary, the proof is complete.   □

Recall that B is a fixed neighborhood of M in X containing the tubular neighborhood $\Theta(X^u(\epsilon_0) \oplus X^s(\epsilon_0))$. We consider a C^1 perturbation $\tilde{T}$ of the time-t map T^t. We set

$$\|\tilde{T} - T^t\|_0 = \sup_{x \in B} |\tilde{T}(x) - T^t(x)|$$

and

$$\|\tilde{T} - T^t\|_1 = \sup_{x \in B} |\tilde{T}(x) - T^t(x)| + \sup_{x \in B} \|D\tilde{T}(x) - DT^t(x)\|$$

A direct consequence of Lemma 5.2 is

Lemma 5.3. *Let $\lambda_2 \in (\lambda_1, 1)$ and let μ, ϵ, and δ be the same as in Lemma 5.2. Then there exists a $\sigma > 0$ such that if*

$$\|\tilde{T} - T^t\|_1 < \sigma,$$

we have

  (i) $\tilde{T}\big(K_g(\epsilon, \mu, \delta)\big) \cap \Theta(\overline{X^u(\epsilon)} \oplus \overline{X^s(\epsilon)}) \subset K_g(\epsilon, \lambda_2^2\mu, \lambda_2^{-1}\delta),$
  (ii) $\tilde{T}\big(K_f(\epsilon, \mu, \delta)\big) \cap \Theta(\overline{X^u(\epsilon)} \oplus \overline{X^s(\epsilon)}) \subset K_f(\epsilon, \lambda_2^2\mu, \lambda_2\delta).$

The invariance of the gap cone $K_g(\epsilon, \mu, \delta)$ and the fattened cone $K_f(\epsilon, \mu, \delta)$ is preserved under small C^1 perturbation of the time-t map, however the invariance of the regular cone $K(\epsilon, \mu)$ is not preserved under the perturbation since M will no longer be invariant. This is the reason we introduce the gap and fattened cones.

Finally, we establish the invariance of moving cones, which will be stated in terms of inequalities.

Before we state this moving cone lemma, we need a lemma which is a generalization of Lemma 5.1.

Lemma 5.4. *For $\epsilon_2 < \epsilon_0$, there exist positive constants ϵ^*, $\tilde{\epsilon}^*$ and σ such that if $\epsilon < \epsilon^*$, $\tilde{\epsilon} < \tilde{\epsilon}^*$ and $\|\tilde{T} - T^t\|_0 < \sigma$, then*

$$\tilde{T}\big(\Theta(\overline{X^u(\epsilon)} \oplus \overline{X^s(\epsilon)})\big) \subset \Theta(X^u(\epsilon_2) \oplus X^s(\epsilon_2))$$

and for all $(m, x^u + x^s) \in \overline{X^u(\epsilon)} \oplus \overline{X^s(\epsilon)}$ *and* $\hat{x}^\alpha \in \overline{X_m^\alpha(\tilde{\epsilon})}, \alpha = u, s, c,$

$$m + x^u + x^s + \hat{x}^u + \hat{x}^s + \hat{x}^c \in \Theta(X^u(\epsilon_2) \oplus X^s(\epsilon_2)) \tag{5.11}$$

and

$$\tilde{T}(m + x^u + x^s + \hat{x}^u + \hat{x}^s + \hat{x}^c) \in \Theta(X^u(\epsilon_2) \oplus X^s(\epsilon_2)). \tag{5.12}$$

Again the proof of this lemma, which we omit, is based on the compactness of M and the continuity of T^t and $\tilde{T}$. Now, for each $(m, x^u + x^s) \in \overline{X^u(\epsilon)} \oplus \overline{X^s(\epsilon)}$ we may write $\tilde{T}(m + x^u + x^s)$ as

$$\tilde{T}(m + x^u + x^s) = m_1 + x_1^u + x_1^s$$

where $x_1^\alpha \in X_{m_1}^\alpha(\epsilon_2), \alpha = u, s$. We may also write

$$\tilde{T}(m + x^u + x^s + \hat{x}^u + \hat{x}^s + \hat{x}^c) = m_1 + x_1^u + x_1^s + \hat{x}_1^u + \hat{x}_1^s + \hat{x}_1^c. \tag{5.13}$$

where $\hat{x}_1^\alpha \in X_m^\alpha, \alpha = u, s, c$.

The next results state the "invariance" of moving cones in terms of inequalities.

Lemma 5.5. *There exist positive constants* ϵ^*, $\tilde{\epsilon}^*$ *and* σ *such that if* $\epsilon < \epsilon^*$, $\tilde{\epsilon} < \tilde{\epsilon}^*$ *and* $\|\tilde{T} - T^t\|_1 < \sigma$ *then the following statements hold for all* $(m, x^u + x^s) \in \overline{X^u(\epsilon)} \oplus$ $\overline{X^s(\epsilon)}, \hat{x}^\alpha \in \overline{X_m^\alpha(\tilde{\epsilon})}, \ \alpha = u, s, c:$

(i) *if* $|\hat{x}^s| \leq \mu(|\hat{x}^u| + |\hat{x}^c|)$ *then* $|\hat{x}_1^s| \leq \lambda_1 \mu(|\hat{x}_1^u| + |\hat{x}_1^c|),$
(ii) *if* $|\hat{x}^s| + |\hat{x}^c| \leq \mu|\hat{x}^u|$ *then* $|\hat{x}_1^s| + |\hat{x}_1^c| \leq \lambda_1 \mu|\hat{x}_1^u|.$

Remark. *If we define the following local cones, this lemma says that they are invariant under* $\tilde{T}$ *relative to the evolving vertex (See Figure 3). For* $(m, x^u + x^s) \in$ $X^u(\epsilon) \oplus X^s(\epsilon)$, *and for* $\mu > 0$, *define*

$$K_u(m + x^u + x^s, \mu, \tilde{\epsilon})$$
$$= \Big\{ m + x^u + x^s + \tilde{x}^u + \tilde{x}^s + \tilde{x}^c \ :$$
$$\tilde{x}^\alpha \in X_m^\alpha(\tilde{\epsilon}) \ for \ \alpha = u, s, \ and \ c, \mu|\tilde{x}^u| > (|\tilde{x}^c| + |\tilde{x}^s|) \Big\}$$

and

$$K_{cu}(m + x^u + x^s, \mu, \tilde{\epsilon})$$
$$= \Big\{ m + x^u + x^s + \tilde{x}^u + \tilde{x}^s + \tilde{x}^c \ :$$
$$\tilde{x}^\alpha \in X_m^\alpha(\tilde{\epsilon}) \ for \ \alpha = u, s, \ and \ c, \mu(|\tilde{x}^u| + |\tilde{x}^c|) > |\tilde{x}^s| \Big\}.$$

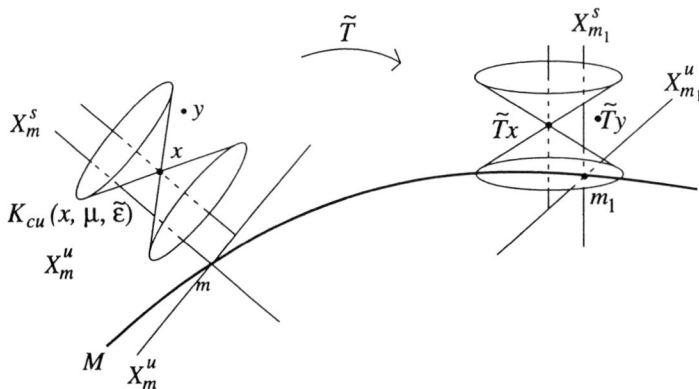

Figure 3. Moving Local Cones

Proof of Lemma 5.5. Applying the Taylor expansion to (5.13), we have

$$
\begin{aligned}
m_1 + x_1^u &+ x_1^s + \hat{x}_1^u + \hat{x}_1^s + \hat{x}_1^c \\
&= \tilde{T}(m + x^u + x^s) + D\tilde{T}(m + x^u + x^s)(\hat{x}^u + \hat{x}^s + \hat{x}^c) \\
&\quad + O(\tilde{\epsilon} + \epsilon + \sigma)|\hat{x}^u + \hat{x}^s + \hat{x}^c|,
\end{aligned}
\tag{5.14}
$$

where $O(\tilde{\epsilon} + \epsilon + \sigma) \to 0$ as $\tilde{\epsilon} + \epsilon + \sigma \to 0$ uniformly in $m \in M$. From (5.14) and the continuity of DT^t it follows that

$$
\begin{aligned}
m_1 + x_1^u &+ x_1^s + \hat{x}_1^u + \hat{x}_1^s + \hat{x}_1^c \\
&= m_1 + x_1^u + x_1^s + DT^t(m)(\hat{x}^u + \hat{x}^s + \hat{x}^c) \\
&\quad + (D\tilde{T} - DT^t)(m + x^u + x^s)(\hat{x}^u + \hat{x}^s + \hat{x}^c) + O(\tilde{\epsilon} + \epsilon + \sigma)|\hat{x}^u + \hat{x}^s + \hat{x}^c|.
\end{aligned}
\tag{5.15}
$$

Similarly,

$$
\begin{aligned}
m_1 + x_1^u &+ x_1^s \\
&= T^t(m) + DT^t(m)(x^u + x^s) + (\tilde{T}(m) - T^t(m)) \\
&\quad + (D\tilde{T}(m) - DT^t(m))(x^u + x^s) + O(\epsilon + \sigma)|x^u + x^s|.
\end{aligned}
\tag{5.16}
$$

Thus,

$$
\begin{aligned}
|m_1 - T^t(m)| &\le |x_1^u| + |x_1^s| + \|DT^t(m)\| |x^u + x^s| \\
&\quad + \|\tilde{T} - T^t\|_1 (1 + |x^u + x^s|) + O(\epsilon + \sigma)|x^u + x^s| \\
&\le 2\epsilon_2 + (C + O(\epsilon + \sigma))\epsilon + \sigma(1 + 2\epsilon).
\end{aligned}
$$

As in the proof of Lemma 5.2, letting $\epsilon_1 < \frac{1}{2C}$, by Lemma 4.4 there exists $\beta^* > 0$ such that (4.12) holds with ϵ_1 instead of ϵ. Let η and r satisfy (4.14). Choose ϵ_2 such that $\epsilon_2 \leq \frac{1}{4}r$ and choose ϵ^* and σ such that

$$(C + O(\epsilon + \sigma))\epsilon + \sigma(1 + 2\epsilon) \leq \frac{r}{2},$$

which implies that m_1 is in an η-neighborhood of $T^t(m)$. For simplicity, we denote $T^t(m)$ by $\bar{m}$. Applying Lemma 4.5 to points $\bar{m}$ and $m_1 + x_1^u + x_1^s$ and using (5.16), we obtain

$$
\begin{aligned}
&|m_1 - T^t(m)| \\
&\leq \frac{1}{1 - C\epsilon_1}|\Pi_{\bar{m}}^c(\tilde{T}(m) - T^t(m) \\
&\quad + (D\tilde{T}(m) - DT^t(m))(x^u + x^s) + O(\epsilon + \sigma)(|x^u + x^s|))| \\
&\leq 2(C\|\tilde{T} - T^t\|_0 + (\|\tilde{T} - T^t\|_1 + O(\epsilon + \sigma))|x^u + x^s|).
\end{aligned}
$$
(5.17)

Applying the projection $\Pi_{m_1}^\alpha$ to (5.15), we find

$$
\begin{aligned}
\hat{x}_1^\alpha = {}& DT^t(m)\hat{x}^\alpha + (\Pi_{m_1}^\alpha - \Pi_{\bar{m}}^\alpha)DT^t(m)\hat{x}^\alpha \\
& + \Pi_{m_1}^\alpha(\Pi_{m_1}^u - \Pi_{\bar{m}}^u)DT^t(m)(\hat{x}^s + \hat{x}^c) \\
& + \Pi_{m_1}^\alpha(\Pi_{m_1}^s - \Pi_{\bar{m}}^s)DT^t(m)(\hat{x}^u + \hat{x}^c) \\
& + \Pi_{m_1}^\alpha(\Pi_{m_1}^c - \Pi_{\bar{m}}^c)DT^t(m)(\hat{x}^u + \hat{x}^s) \\
& + \Pi_{m_1}^\alpha(D\tilde{T} - DT^t)(m + x^u + x^s)(\hat{x}^u + \hat{x}^s + \hat{x}^c) \\
& + \Pi_{m_1}^\alpha O(\tilde{\epsilon} + \epsilon + \sigma)|\hat{x}^u + \hat{x}^s + \hat{x}^c|
\end{aligned}
$$
(5.18)

for $\alpha = u, s, c$. Note that

$$\|\Pi_{m_1}^\alpha - \Pi_{\bar{m}}^\alpha\| \leq \text{Lip}_m(\Pi_m^\alpha)|m_1 - \bar{m}|.$$
(5.19)

Thus, from (5.17) and (5.18) there exists a positive constant C, which does not depend on m, ϵ, $\tilde{\epsilon}$, $\tilde{T}$, $\hat{x}_1^\alpha$ and $\hat{x}^\alpha$, such that

$$
\begin{aligned}
&|\hat{x}_1^u| \geq |DT^t(m)\hat{x}^u| - C(\|\tilde{T} - T^t\|_1 + O(\tilde{\epsilon} + \epsilon + \sigma))(|\hat{x}^u| + |\hat{x}^s| + |\hat{x}^c|), \\
&|\hat{x}_1^s| \leq |DT^t(m)\hat{x}^s| + C(\|\tilde{T} - T^t\|_1 + O(\tilde{\epsilon} + \epsilon + \sigma))(|\hat{x}^u| + |\hat{x}^s| + |\hat{x}^c|), \\
&|\hat{x}_1^c| \geq |DT^t(m)\hat{x}^c| - C(\|\tilde{T} - T^t\|_1 + O(\tilde{\epsilon} + \epsilon + \sigma))(|\hat{x}^u| + |\hat{x}^s| + |\hat{x}^c|).
\end{aligned}
$$

In the following, as in Section 4, C is a generic constant whose value may change from line to line, but it depends only on the projections and the time-t map T^t.

Thus,

$$\lambda_1\mu(|\hat{x}_1^u| + |\hat{x}_1^c|) - |\hat{x}_1^s|$$

$$\geq \lambda_1\mu\left(\inf_{x^u\in X_m^u, |x^u|=1}|DT^t(m)x^u||\hat{x}^u| + \inf_{x^c\in X_m^c, |x^c|=1}|DT^t(m)x^c||\hat{x}^c|\right)$$

$$- \|DT^t(m)|_{X_m^s}\| \, |\hat{x}^s| - (\lambda_1\mu + 1)C(\|\tilde{T} - T^t\|_1 + O(\tilde{\epsilon} + \epsilon + \sigma))(|\hat{x}^u| + |\hat{x}^s| + |\hat{x}^c|)$$

which, by (H3), is larger than

$$\inf_{|x^c|=1}|DT^t(m)x^c|\left(\frac{\lambda_1\mu}{\lambda}|\hat{x}^u| + \lambda_1\mu|\hat{x}^c| - \lambda|\hat{x}^s|\right)$$

$$- (\lambda_1\mu + 1)C(\|\tilde{T} - T^t\|_1 + O(\tilde{\epsilon} + \epsilon + \sigma))(|\hat{x}^u| + |\hat{x}^s| + |\hat{x}^c|)$$

$$\geq \inf_{|x^c|=1}|DT^t(m)x^c|\left(\frac{\lambda_1\mu}{\lambda}|\hat{x}^u| + \lambda_1\mu|\hat{x}^c| - \lambda\mu(|\hat{x}^u| + |\hat{x}^c|)\right)$$

$$- (\lambda_1\mu + 1)C(\|\tilde{T} - T^t\|_1 + O(\tilde{\epsilon} + \epsilon + \sigma))(|\hat{x}^u| + |\hat{x}^s| + |\hat{x}^c|)$$

$$\geq \left(\mu(\lambda_1 - \lambda)\inf_{|x^c|=1}|DT^t(m)x^c| - (\lambda_1\mu + 1)C(\|\tilde{T} - T^t\|_1 + O(\tilde{\epsilon} + \epsilon + \sigma))(1+\mu)\right)$$

$$(|\hat{x}^u| + |\hat{x}^c|),$$

where the assumption in (i) of this lemma was used. Since $\inf_{|x^c|=1}|DT^t(m)x^c| > 0$ and $\lambda < \lambda_1 < 1$, one may choose ϵ^*, $\tilde{\epsilon}^*$ and σ sufficiently small so that

$$\mu(\lambda_1 - \lambda)\inf_{|x^c|=1}|DT^t(m)x^c| - (\lambda_1\mu + 1)C(\|\tilde{T} - T^t\|_1 + O(\tilde{\epsilon} + \epsilon + \sigma))(1+\mu) > 0.$$

This proves (i).

From (5.18) we also obtain

$$|\hat{x}_1^u| \geq |DT^t(m)\hat{x}^u| - C(\|\tilde{T} - T^t\|_1 + O(\tilde{\epsilon} + \epsilon + \sigma))(|\hat{x}^u| + |\hat{x}^s| + |\hat{x}^c|)$$

and

$$|\hat{x}_1^\alpha| \leq |DT^t(m)\hat{x}^\alpha| + C(\|\tilde{T} - T^t\|_1 + O(\tilde{\epsilon} + \epsilon + \sigma))(|\hat{x}^u| + |\hat{x}^s| + |\hat{x}^c|)$$

for $\alpha = s, c$. Thus, by using the normal hyperbolicity and the assumption in (ii) of this lemma, we obtain

$$\lambda_1 \mu |\hat{x}_1^u| - (|\hat{x}_1^s| + |\hat{x}_1^c|)$$

$$\geq \|DT^t(m)|_{X_m^c}\|(\frac{\lambda_1}{\lambda}\mu|\hat{x}^u| - \lambda|\hat{x}^s| - |\hat{x}^c|)$$

$$- (\lambda_1\mu + 1)C(\|\tilde{T} - T^t\|_1 + O(\tilde{\epsilon} + \epsilon + \sigma))(|\hat{x}^u| + |\hat{x}^s| + |\hat{x}^c|)$$

$$\geq \|DT^t(m)|_{X_m^c}\| \left(\frac{\lambda_1}{\lambda}(|\hat{x}^s| + |\hat{x}^c|) - \lambda|\hat{x}^s| - |\hat{x}^c| \right)$$

$$- (\lambda_1\mu + 1)C(\|\tilde{T} - T^t\|_1 + O(\tilde{\epsilon} + \epsilon + \sigma))(|\hat{x}^u| + |\hat{x}^s| + |\hat{x}^c|)$$

$$\geq \left(\|DT^t(m)|_{X_m^c}\| \left(\frac{\lambda_1}{\lambda} - 1 \right) - (\lambda_1\mu + 1)C(\|\tilde{T} - T^t\|_1 + O(\tilde{\epsilon} + \epsilon + \sigma))(1 + \mu) \right) |\hat{x}^s| + |\hat{x}^c|$$

$$\geq 0,$$

which is achieved by choosing ϵ^*, $\tilde{\epsilon}$ and σ smaller if necessary. This completes the proof. $\qquad\square$

6. Center-Unstable Manifold.

We now have the machinery to construct the manifolds given in Theorems A and B, starting with the local center-unstable manifold (see Section 3 for the definition). The idea is to take Lipschitz graphs over $X^u(\epsilon)$ in $X^u(\epsilon) \oplus X^s(\epsilon)$ and map them under $\tilde{T}$, the perturbation of T^t for some $t > t_0$. Using the cone lemmas we show that in a certain sense, this "graph transform" is a contraction on the space of Lipschitz graphs. The fixed point is the desired invariant manifold.

We regard the normal bundle $X^u \oplus X^s$ as a bundle over the unstable bundle X^u and consider sections of this bundle in the tubular neighborhood. In particular, we define the complete metric space Γ^{cu} of Lipschitz sections, which are described in terms of the global fattened cone and local cones. The key to construct a "graph transform" in the space Γ^{cu} is to show that for any given Lipschitz section $h \in \Gamma^{cu}$, the image of its graph under $\tilde{T}$ restricted to the tubular neighborhood is the graph of a Lipschitz section in Γ^{cu}. This is given in Proposition 6.4 and yields a graph transform $\mathcal{F}^{cu}$ defined on Γ^{cu}. Using the invariance of the cones, one can prove that $\mathcal{F}^{cu}$ is a contraction and the fixed point is the desired center-unstable manifold. Because we do not have a global Cartesian coordinate system, we must interpret the Lipschitz property in the various local coordinates. This is the point of the technical lemmas 6.1 and 6.1. Another difficulty is that T^t is not a diffeomorphism (not even a homeomorphism, in fact neither one-to-one nor onto) and therefore Lipschitz graphs are not automatically mapped into Lipschitz graphs. Estimates due to the normal hyperbolicity are used to overcome this difficulty, and allow us to define the graph transform.

Throughout this paper, we shall encounter several positive parameters: ϵ – the size of a tubular neighborhood; $\tilde{\epsilon}$ – the size of the perturbation of the tubular neighborhood to accommodate local cones; δ – the size of the neck of the fat and gap cones; σ – the C^1 size of the perturbation from the time-t map T^t. We use ϵ^*, $\tilde{\epsilon}^*$ and δ^* to denote the upper bounds for the parameters ϵ, $\tilde{\epsilon}$ and δ, respectively. The values of ϵ^*, $\tilde{\epsilon}^*$ and δ^* may change from lemma to lemma, but they are equal or smaller than those in the previous related lemmas. Without loss of generality, we assume that $\max\{\epsilon^*, \tilde{\epsilon}^*, \delta^*\} \leq 1$. As before, we will use C for a generic constant whose value may change from line to line, but it depends only on the projections and the time-t map. The quantity $O(\epsilon)$ also depends only the projections and the time-t map.

We shall discuss sections of this bundle, $h : X^u(\epsilon) \to X^u(\epsilon) \oplus X^s(\epsilon)$. With an obvious abuse of notation, for such a section we will usually write $h(m, x^u)$ for the point in $X_m^s(\epsilon)$ rather than the point in $X_m^u(\epsilon) \oplus X_m^s(\epsilon)$ which is the image of (m, x^u) under h. By the "graph" of the section h we shall mean

$$gr(h) \equiv \{(m, x^u + x^s) \in X^u(\epsilon) \oplus X^s(\epsilon) : x^s = h(m, x^u)\}.$$

Define, for fixed $\mu \in (0,1)$, $\epsilon > 0$, $0 < \delta < \epsilon$, and $\tilde{\epsilon} > 0$ the space of sections

$$\Gamma^{cu} = \Gamma^{cu}(\epsilon, \mu, \delta, \tilde{\epsilon})$$
$$\equiv \Big\{ h : \overline{X^u(\epsilon)} \to \overline{X^u(\epsilon)} \oplus \overline{X^s(\epsilon)} : gr(h) \subset K_f(\epsilon, \mu, \delta)$$
$$\text{and for all } (m, x^u) \in \overline{X^u(\epsilon)}$$
$$\text{if } \Theta^{-1}(m + x^u + h(m, x^u) + x_1^u + x_1^s + x_1^c) \in gr(h)$$
$$\text{for } x_1^\alpha \in \overline{X_m^\alpha(\tilde{\epsilon})}, \text{ then } |x_1^s| \le \mu(|x_1^u| + |x_1^c|) \Big\}.$$

We may define a norm on Γ^{cu} by

$$\|h\| = \sup \Big\{ |h(m, x^u)| : (m, x^u) \in \overline{X^u(\epsilon)} \Big\},$$

which makes Γ^{cu} into a complete metric space. Observe that from Lemma 4.5, there exist positive constants ϵ^* and $\tilde{\epsilon}^*$ such that if $\epsilon < \epsilon^*$ and $\tilde{\epsilon} < \tilde{\epsilon}^*$ then Γ^{cu} is not empty for $\delta \in (0, \epsilon)$, in fact the zero section is in Γ^{cu}. In what follows we take such a value of $\tilde{\epsilon}^*$ but we will take $\tilde{\epsilon}^*$ possibly smaller than that indicated here.

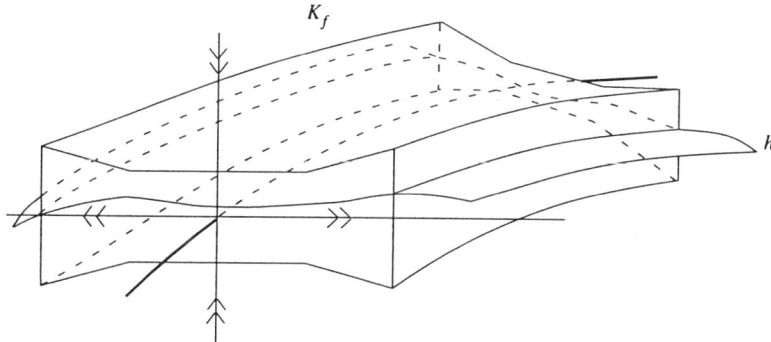

Figure 4. A Lipschitz Section, $h \in \Gamma^{cu}$

Lipschitz sections $h \in \Gamma^{cu}$ induce mappings in the local coordinate systems near M. The next two lemmas give the properties of these induced mappings.

Lemma 6.1. *Let $\mu \in (0,1)$ and $\rho \in (1,2)$. For each $\tilde{\epsilon} < \tilde{\epsilon}^*$ there exists $\epsilon^* > 0$ such that for each $\epsilon \in (0, \epsilon^*)$, $\delta \in (0, \epsilon)$ and for all $h \in \Gamma^{cu}$ and $m_0 \in M$, if*

$$g : (M \cap B(m_0, \rho\epsilon)) \times X_{m_0}^u(\rho^{-1}\epsilon) \to X_{m_0}^s$$

is defined by

$$\Pi_m^s g(m, \tilde{x}^u) = h(m, \Pi_m^u \tilde{x}^u), \tag{6.1}$$

then

$$|g(m_1, \tilde{x}_1^u) - g(m_2, \tilde{x}_2^u)| \leq \rho\mu \left(|m_1 - m_2| + |\tilde{x}_1^u - \tilde{x}_2^u|\right). \qquad (6.2)$$

Proof. Let $\epsilon_1 < \epsilon_0$, where ϵ_0 is introduced in Section 5. By Lemma 4.4, there exists $\beta^* > 0$ such that (4.12) holds with ϵ_1 instead of ϵ. Let η and r satisfy (4.14) and $\eta < \rho - 1$. For $m_0 \in M$ choose $\epsilon^* < \min\{\epsilon_1, r/2\}$. For $\epsilon < \epsilon^*$ by (4.5) $\Pi_m^u X_{m_0}^u(\rho^{-1}\epsilon) \subset X_m^u(\epsilon)$ for $m \in B(m_0, \rho\epsilon)$ and $\Pi_m^s|_{X_{m_0}^s}$ is an isomorphism. Thus g is well-defined on $(M \cap B(m_0, \rho\epsilon)) \times X_{m_0}^u(\rho^{-1}\epsilon)$.

To show that (6.2) holds, for $i = 1, 2$ let $m_i \in B(m_0, \rho\epsilon), \tilde{x}_i^u \in X_{m_0}^u(\rho^{-1}\epsilon)$ and $\tilde{x}_i^s = g(m_i, \tilde{x}_i^u)$. We represent points in different coordinate systems: with $\bar{x}_i^\alpha \in X_{m_0}^\alpha$, for $\alpha = u, s, c$

$$\begin{aligned} m_i + \Pi_{m_i}^u \tilde{x}_i^u + \Pi_{m_i}^s \tilde{x}_i^s &= m_0 + \bar{x}_i^u + \bar{x}_i^s + \bar{x}_i^c \\ &= m_i + x_i^u + x_i^s, \end{aligned} \qquad (6.3)$$

where $x_i^u = \Pi_{m_i}^u \tilde{x}_i^u$ and $x_i^s = \Pi_{m_i}^s \tilde{x}_i^s = \Pi_{m_i}^s g(m_i, \tilde{x}_i^u) = h(m_i, \Pi_{m_i}^u \tilde{x}_i^u)$.

Since $m_i \in B(m_0, \rho\epsilon)$ and $\tilde{x}_i^u \in X_{m_0}^u(\rho^{-1}\epsilon)$ we have $|x_i^u| < \epsilon$ and so $|x_i^s| = |h(m_i, \Pi_{m_i}^u \tilde{x}_i^u)| \leq \epsilon$ since $h \in \Gamma^{cu}$. By (4.5) and (6.1) it follows that $|\tilde{x}_i^s| = |g(m_i, \tilde{x}_i^u)| \leq \frac{\epsilon}{1-\eta} < 2\epsilon$.

From (6.3) we have $|\bar{x}_i^u + \bar{x}_i^s + \bar{x}_i^c| \leq |m_0 - m_i| + |x_i^u| + |x_i^s| < 4\epsilon$ and so $|\bar{x}_i^\alpha| \leq C\epsilon$ for some constant C, for $\alpha = u, s, c$. By Lemma 4.5, for a possibly larger value of C

$$|\tilde{x}_1^s - \tilde{x}_2^s - (\bar{x}_1^s - \bar{x}_2^s)| \leq C\epsilon_1|m_1 - m_2| + C|m_1 - m_0| \left(|\tilde{x}_1^u - \tilde{x}_2^u| + |\tilde{x}_1^s - \tilde{x}_2^s|\right)$$

and so

$$(1 - 2C\epsilon)|\tilde{x}_1^s - \tilde{x}_2^s| \leq C\epsilon_1|m_1 - m_2| + 2C\epsilon|\tilde{x}_1^u - \tilde{x}_2^u| + |\bar{x}_1^s - \bar{x}_2^s|. \qquad (6.4)$$

To obtain (6.2) we must estimate $|\bar{x}_1^s - \bar{x}_2^s|$. First we use (4.15) to write

$$m_2 + x_2^u + x_2^s = m_1 + \hat{x}_2^u + \hat{x}_2^s + \hat{x}_2^c$$

where $\hat{x}_2^\alpha \in X_{m_1}^\alpha$ for $\alpha = u, s, c$. Note that for some constant $C > 0$ $|\hat{x}_2^\alpha| \leq \|\Pi_{m_1}^\alpha\| |m_2 - m_1 + x_2^u + x_2^s| \leq C\epsilon$.

Also,

$$m_1 + x_1^u + h(m_1, x_1^u) + (\hat{x}_2^u - x_1^u) + (\hat{x}_2^s - x_1^s) + \hat{x}_2^c$$

$$= m_2 + x_2^u + x_2^s = m_2 + x_2^u + h(m_2, x_2^u) \in \Theta(gr(h)).$$

By choosing ϵ^* sufficiently small, we find that $|\hat{x}_2^u - x_1^u| \leq C\epsilon < \tilde{\epsilon}, |\hat{x}_2^s - x_1^s| \leq C\epsilon < \tilde{\epsilon}$ and $|\hat{x}_2^c| \leq C\epsilon < \tilde{\epsilon}$, hence from the fact that $h \in \Gamma^{cu}$,

$$|\hat{x}_2^s - x_1^s| \leq \mu \left(|\hat{x}_2^u - x_1^u| + |\hat{x}_2^c|\right). \qquad (6.5)$$

Now

$$\bar{x}_1^s - \bar{x}_2^s = \Pi_{m_0}^s \left(m_1 + x_1^u + x_1^s - (m_2 + x_2^u + x_2^s) \right)$$
$$= \Pi_{m_0}^s \left(x_1^u - \hat{x}_2^u + x_1^s - \hat{x}_2^s - \hat{x}_2^c \right)$$

so, by (4.5)

$$|\bar{x}_1^s - \bar{x}_2^s| \leq |\left(\Pi_{m_0}^s - \Pi_{m_1}^s \right) \left(x_1^u - \hat{x}_2^u - \hat{x}_2^c \right)| + (1+\eta)|x_1^s - \hat{x}_2^s|$$

and (6.5) with (H4) gives

$$|\bar{x}_1^s - \bar{x}_2^s| \leq (C|m_0 - m_1| + (1+\eta)\mu)(|x_1^u - \hat{x}_2^u| + |\hat{x}_2^c|)$$
$$\leq (C\epsilon + (1+\eta)\mu)(|x_1^u - \hat{x}_2^u| + |\hat{x}_2^c|). \tag{6.6}$$

Using (6.3) and the definition of $\hat{x}_2^u$ we have

$$|x_1^u - \hat{x}_2^u| = |\Pi_{m_1}^u \left(\bar{x}_1^u - \bar{x}_2^u + \bar{x}_1^s - \bar{x}_2^s + \bar{x}_1^c - \bar{x}_2^c \right)|$$
$$\leq \|\Pi_{m_1}^u |_{X_{m_0}^u} \| |\bar{x}_1^u - \bar{x}_2^u| + \|\Pi_{m_1}^u - \Pi_{m_0}^u\| \left(|\bar{x}_1^s - \bar{x}_2^s| + |\bar{x}_1^c - \bar{x}_2^c| \right)$$
$$\leq (1+\eta)|\bar{x}_1^u - \bar{x}_2^u| + C\epsilon \left(|\bar{x}_1^s - \bar{x}_2^s) + |\bar{x}_1^c - \bar{x}_2^c| \right). \tag{6.7}$$

Similarly,

$$|\hat{x}_2^c| \leq (1+\eta)|\bar{x}_1^c - \bar{x}_2^c| + C\epsilon \left(|\bar{x}_1^s - \bar{x}_2^s| + |\bar{x}_1^u - \bar{x}_2^u| \right). \tag{6.8}$$

Combining (6.6)–(6.8) gives

$$|\bar{x}_1^s - \bar{x}_2^s| \leq (O(\epsilon) + (1+\eta)^2\mu)(|\bar{x}_1^u - \bar{x}_2^u| + |\bar{x}_1^c - \bar{x}_2^c|). \tag{6.9}$$

Now (4.16) and (4.17) may be used for (6.3) to get, respectively,

$$|\bar{x}_1^c - \bar{x}_2^c| \leq (1 + C\epsilon_1)|m_1 - m_2| + C\epsilon(|\tilde{x}_1^u - \tilde{x}_2^u| + |\tilde{x}_1^s - \tilde{x}_2^s|),$$

and

$$|\bar{x}_1^u - \bar{x}_2^u| \leq C\epsilon_1|m_1 - m_2| + (1 + C\epsilon)|\tilde{x}_1^u - \tilde{x}_2^u| + C\epsilon|\tilde{x}_1^s - \tilde{x}_2^s|,$$

which, with (6.9) give

$$|\bar{x}_1^s - \bar{x}_2^s|$$
$$\leq \left(O(\epsilon) + (1+\eta)^2\mu \right)\left[(1 + C\epsilon_1)|m_1 - m_2| + (1 + C\epsilon)|\tilde{x}_1^u - \tilde{x}_2^u| + C\epsilon|\tilde{x}_1^s - \tilde{x}_2^s| \right].$$

Using this in (6.4) yields,

$$|\tilde{x}_1^s - \tilde{x}_2^s|$$
$$\leq \left((1+\eta)^2\mu + O(\epsilon) + O(\epsilon_1) \right)\left(|m_1 - m_2| + |\tilde{x}_1^u - \tilde{x}_2^u| \right).$$

Let η also satisfy that $\eta < \sqrt{\frac{1+\rho}{2}} - 1$. Then one may choose ϵ_1 and ϵ^* small enough such that

$$(1+\eta)^2 \mu + O(\epsilon) + O(\epsilon_1) < \rho\mu$$

for $\epsilon < \epsilon^*$. Thus

$$\begin{aligned}
&|g(m_1, \tilde{x}_1^u) - g(m_2, \tilde{x}_2^u)| \\
&= |\tilde{x}_1^s - \tilde{x}_2^s| \\
&\leq \rho\mu(|m_1 - m_2| + |\tilde{x}_1^u - \tilde{x}_2^u|).
\end{aligned}$$

The proof is complete.   □

The next lemma is similar but set in a local Cartesian coordinate system. Here $\tilde{\epsilon}^*$ is as before.

Lemma 6.2. *Let $\mu \in (0,1)$ and $\rho \in (1,2)$. Then for each $\tilde{\epsilon} < \tilde{\epsilon}^*$ there exists $\epsilon^* = \epsilon^*(\tilde{\epsilon})$ and $\delta^* = \delta^*(\epsilon) < \epsilon$ such that for each $\epsilon < \epsilon^*$ and $\delta < \delta^*$ if $h \in \Gamma^{cu}(\epsilon, \mu, \delta, \tilde{\epsilon})$ and $m_0 \in M$, then there exists*

$$f : \overline{X_{m_0}^u(\rho^{-2}\epsilon)} \times \overline{X_{m_0}^c(\epsilon)} \to X_{m_0}^s(\epsilon)$$

such that

$$|f(x_1^u, x_1^c) - f(x_2^u, x_2^c)| \leq \rho\mu(|x_1^u - x_2^u| + |x_1^c - x_2^u|)$$

and

$$\Theta^{-1}(m_0 + x_0^c + x_0^u + f(x_0^u, x_0^c)) \in gr(h) \text{ for all } (x_0^u, x_0^c) \in \overline{X_{m_0}^u(\rho^{-2}\epsilon)} \times \overline{X_{m_0}^c(\epsilon)}.$$

Furthermore, for all $m \in M \cap B(m_0, \rho^{-1}\epsilon)$ and $x^u \in \overline{X_n^u(\rho^{-3}\epsilon)}$

$$m + x^u + h(m, x^u) = m_0 + x_0^c + x_0^u + f(x_0^u, x_0^c)$$

for some $(x_0^c, x_0^u) \in \overline{X_{m_0}^u(\rho^{-2}\epsilon)} \times \overline{X_{m_0}^c(\epsilon)}$.

Proof. Fix ϵ_1, η and r to satisfy the requirements in the proof of Lemma 6.1. Take $\delta \in (0, \epsilon)$ fixed but restrict both δ and ϵ_1 to satisfy further conditions to be specified later. Take ϵ^* as given by Lemma 6.1 and also require $\epsilon^* < \epsilon_1$. Choose $\epsilon \leq \epsilon^*$. Thus we may apply Lemma 6.1 and some estimates obtained in its proof to establish this lemma.

To construct f for $h \in \Gamma^{cu}$ we must show that for all $(x_0^u, x_0^c) \in \overline{X_{m_0}^u(\rho^{-2}\epsilon)} \times \overline{X_{m_0}^c(\epsilon)}$ there exists a unique $x^s \in X_{m_0}^s(\epsilon)$ such that $m_0 + x_0^c + x_0^u + x^s \in \Theta(gr(h))$.

Define a map $\xi : \overline{X_{m_0}^s(\epsilon)} \to X_{m_0}^s$ as follows:

For $0 < \epsilon_2 < \epsilon_1$, by Lemma 5.4, we may choose a ϵ^* sufficiently small such that for $x^\alpha \in \overline{X_{m_0}^\alpha(\epsilon)}$ for $\alpha = u, s, c$ and $\epsilon < \epsilon^*$, $\Theta^{-1}(m_0 + x^u + x^s + x^c) \in X^u(\epsilon_2) \oplus X^s(\epsilon_2)$.

Thus, for each $x^s \in \overline{X^s_{m_0}(\epsilon)}$ there exists a unique point $(m, \bar{x}^u + \bar{x}^s)$ in $X^u(\epsilon_2) \oplus X^s(\epsilon_2)$ such that

$$m_0 + x^c_0 + x^u_0 + x^s = m + \bar{x}^u + \bar{x}^s \tag{6.10}$$

and

$$|m - m_0| \le 5\epsilon_2.$$

Choosing ϵ_2 small enough, m is in the η-neighborhood of m_0 and so we may apply Lemma 4.5. Taking the two points as $m + \bar{x}^u + \bar{x}^s$ and m_0 in (4.16). (4.17) gives

$$|m - m_0| \le (1 - C\epsilon_1)^{-1} |x^c_0| < 2\epsilon \tag{6.11}$$

and

$$|\bar{x}^u| \le (1 + \eta) \left| \left(\Pi^u_m |_{X^u_{m_0}} \right)^{-1} \bar{x}^u \right| \le (1 + \eta)(|x^u_0| + C\epsilon_1 |m - m_0|) < \rho^{-1}\epsilon$$

provided that ϵ_1 satisfies $\epsilon_1 < \frac{\rho - 1}{2C(\rho + 1)\rho^2}$. Consequently, there exist $\hat{x}^\alpha \in X^\alpha_{m_0}, \alpha = u, s, c$, such that

$$m + \bar{x}^u + h(m, \bar{x}^u) = m_0 + \hat{x}^u + \hat{x}^s + \hat{x}^c.$$

Define

$$\xi(x^s) = \hat{x}^s.$$

We claim that ξ is a contraction on $\overline{X^s_{m_0}(\epsilon)}$.

We first show that ξ maps $\overline{X^s_{m_0}(\epsilon)}$ to $X^s_{m_0}(\epsilon)$. Letting $\tilde{x}^u = \left(\Pi^u_m |_{X^u_{m_0}} \right)^{-1} \bar{x}^u$, (6.10) and (4.17) give

$$|\tilde{x}^u| \le |x^u_0| + C\epsilon_1 |m - m_0| \le \rho^{-2}\epsilon + C\epsilon_1\epsilon < \rho^{-1}\epsilon,$$

where the last estimate is obtained by choosing ϵ_1 to satisfy $\epsilon_1 < \frac{\rho - 1}{C\rho^2}$. Thus, by Lemma 6.1, we have

$$m + \Pi^u_m \tilde{x}^u + \Pi^s_m g(m, \tilde{x}^u) = m + \bar{x}^u + h(m, \bar{x}^u) = m_0 + \hat{x}^u + \hat{x}^s + \hat{x}^c. \tag{6.12}$$

Let $\tilde{x}^s = g(m, \tilde{x}^u)$. By (4.17), with (6.12), using m_0 as the second point,

$$|\hat{x}^s - g(m, \tilde{x}^u)| \le C\epsilon_1 |m - m_0|$$

and so, by (6.11)

$$|\hat{x}^s - g(m, \tilde{x}^u)| \le C\epsilon_1\epsilon.$$

Using Lemma 6.1 and the relationship between g and $h \in \Gamma^{cu}(\epsilon, \mu, \delta, \tilde{\epsilon})$,

$$|g(m, \tilde{x}^u)| \le \rho\mu|\tilde{x}^u| + |g(m, 0)| \le \rho\mu|\tilde{x}^u| + C\delta.$$

Hence,

$$|\xi(x^s)| = |\hat{x}^s| \leq \mu\epsilon + C\delta + C\epsilon_1\epsilon < \epsilon,$$

by choosing ϵ_1 and δ such that $\epsilon_1 < \frac{1-\mu}{2C}$ and $\delta < \frac{1-\mu}{2C}\epsilon$. Therefore

$$\xi : \overline{X^s_{m_0}(\epsilon)} \to X^s_{m_0}(\epsilon).$$

To show that at ξ is contractive, for $i = 1, 2$, let $x_i^s \in \bar{X}^s_{m_0}(\epsilon)$ and write

$$m_0 + x_0^c + x_0^u + x_i^s = m_i + \bar{x}_i^u + \bar{x}_i^s, \tag{6.13}$$

$$m_i + \bar{x}_i^u + h(m_i, \bar{x}_i^u) = m_i + \Pi^u_{m_i}\tilde{x}_i^u + \Pi^s_{m_i}g(m_i, \tilde{x}_i^u)$$

$$= m_0 + \hat{x}_i^u + \hat{x}_i^c + \hat{x}_i^s.$$

We have $\hat{x}_i^s = \xi(x_i^s)$ and we may write $g(m_i, \tilde{x}_i^u)$ as $\tilde{x}_i^s$ and $\left(\Pi^s_{m_i}\big|_{X^s_{m_0}}\right)^{-1}\bar{x}_i^s$ as $\tilde{\tilde{x}}_i^s$. Then, using (H4), estimates on the projections, and Lemma 6.1,

$$|\xi(x_1^s) - \xi(x_2^s)|$$
$$= |\hat{x}_1^s - \hat{x}_2^s|$$
$$= |\Pi^s_{m_0}\left(m_1 - m_2 + \left(\Pi^u_{m_1} - \Pi^u_{m_2}\right)\tilde{x}_1^u\right.$$
$$+ \Pi^u_{m_2}(\tilde{x}_1^u - \tilde{x}_2^u) + \left(\Pi^s_{m_1} - \Pi^s_{m_2}\right)\tilde{x}_1^s + \Pi^s_{m_2}(g(m_1, \tilde{x}_1^u) - g(m_2, \tilde{x}_2^u)))|$$
$$\leq C(|m_1 - m_2| + \epsilon|m_1 - m_2|$$
$$+ |(\Pi^s_{m_0} - \Pi^s_{m_2})\Pi^u_{m_2}||\tilde{x}_1^u - \tilde{x}_2^u| + \rho\mu(|m_1 - m_2| + |\tilde{x}_1^u - \tilde{x}_2^u|))$$
$$\leq C(|m_1 - m_2| + |\tilde{x}_1^u - \tilde{x}_2^u|). \tag{6.14}$$

where the second C is larger than the first. We now claim that for another constant, C,

$$|m_1 - m_2| \leq C\epsilon|x_1^s - x_2^s| \tag{6.15}$$

and

$$|\tilde{x}_1^u - \tilde{x}_2^u| \leq C\epsilon|x_1^s - x_2^s|. \tag{6.16}$$

These, together with (6.14) show, by choosing sufficiently small ϵ^*, that ξ is a contraction on $\bar{X}^s_{m_0}(\epsilon)$, and hence has a unique fixed point, x_0^s. By defining $f(x_0^u, x_0^c) = x_0^s$, we have from the definition of ξ that $(x_0^u - \hat{x}^u) + (x_0^c - \hat{x}^c) = \bar{x}^s - h(m, \bar{x}^u)$. Applying $\Pi^s_{m_0}$ and noting that this is an isomorphism on X^s_m shows that $\bar{x}^s = h(m, \bar{x}^u)$ and so (6.10) gives the conclusion of the lemma.

It remains to establish (6.15) and (6.16). In what follows, C will be a generic constant whose value may change from line to line. We point out again that the constant C depends only on the projections.

Applying (4.17) to (6.13), we obtain

$$|m_1 - m_2| \leq C\epsilon_1 |m_1 - m_2| + C|m_1 - m_0| \left(|\tilde{x}_1^u - \tilde{x}_2^u| + |\tilde{\tilde{x}}_1^s - \tilde{\tilde{x}}_2^s| \right).$$

Using (6.11), we have

$$|m_1 - m_2| \leq C\epsilon \left(|\tilde{x}_1^u - \tilde{x}_2^u| + |\tilde{\tilde{x}}_1^s - \tilde{\tilde{x}}_2^s| \right). \tag{6.17}$$

Applying (4.17) to (6.13), we get

$$|\tilde{x}_1^u - \tilde{x}_2^u| \leq C\epsilon_1 |m_1 - m_2| + C|m_1 - m_0| \left(|\tilde{x}_1^u - \tilde{x}_2^u| + |\tilde{\tilde{x}}_1^s - \tilde{\tilde{x}}_2^s| \right)$$

which yields

$$|\tilde{x}_1^u - \tilde{x}_2^u| \leq C\epsilon_1 |m_1 - m_2| + C\epsilon |\tilde{\tilde{x}}_1^s - \tilde{\tilde{x}}_2^s|. \tag{6.18}$$

Combining this with (6.17) gives

$$|m_1 - m_2| \leq C\epsilon |\tilde{\tilde{x}}_1^s - \tilde{\tilde{x}}_2^s|. \tag{6.19}$$

Applying $\Pi_{m_i}^s$ to (6.13) gives, for $i = 1, 2$

$$\Pi_{m_i}^s \tilde{\tilde{x}}_i^s = \Pi_{m_i}^s (m_0 + x_0^u + x_0^c + x_i^s - m_i)$$

and so

$$\tilde{\tilde{x}}_i^s = \left(\Pi_{m_i}^s |_{X_{m_0}^s} \right)^{-1} \Pi_{m_i}^s (m_0 + x_0^u + x_0^c + x_i^s - m_i).$$

Therefore, using (4.11),

$$\begin{aligned}
&|\tilde{\tilde{x}}_1^s - \tilde{\tilde{x}}_2^s| \\
&\leq \left| \left[\left(\Pi_{m_1}^s |_{X_{m_0}^s} \right)^{-1} \Pi_{m_1}^s - \left(\Pi_{m_2}^s |_{X_{m_0}^s} \right)^{-1} \Pi_{m_2}^s \right] (m_0 + x_0^u + x_0^c + x_1^s - m_1) \right| \\
&\quad + \left| \left(\Pi_{m_2}^s |_{X_{m_0}^s} \right)^{-1} \Pi_{m_2}^s (x_1^s - x_2^s - m_1 + m_2) \right| \\
&\leq C \left(|m_1 - m_2| + |x_1^s - x_2^s| \right).
\end{aligned} \tag{6.20}$$

This together with (6.19) establishes (6.15). To prove (6.16), combine (6.18), (6.20) and (6.15).

   We prove the last statement. For $i = 1, 2$ let $m_i \in B(m_0, \rho^{-1}\epsilon) \cap M$ and $x_i^u \in \overline{X_{m_i}^u(\rho^{-3}\epsilon)}$, write

$$m_i + x_i^u + h(m_i, x_i^u) = m_0 + \bar{x}_i^u + \bar{x}_i^s + \bar{x}_i^c = m_i + \Pi_{m_i}^u \tilde{x}_i^u + \Pi_{m_i}^s \tilde{x}_i^s, \tag{6.21}$$

for some $\bar{x}_i^\alpha, \tilde{x}_i^\alpha \in X_{m_0}^\alpha, \alpha = u, s, c$ as in (4.15). Denote $h(m_i, x_i^u)$ by x_i^s, then $|x_i^s| \leq \epsilon$ since $h \in \Gamma^{cu}$.

By (4.5), for $\alpha = u, s$ and $i = 1, 2$

$$|\tilde{x}_i^\alpha| \leq (1 - \eta)^{-1}|\Pi_{m_i}^\alpha \tilde{x}_i^\alpha| = (1 - \eta)^{-1}|x_i^\alpha| \leq 2\epsilon. \tag{6.22}$$

By Lemma 4.5, using m_0 as one of the two points in (4.16) and (4.17),

$$|\bar{x}_i^c| \leq (1 + C\epsilon_1)|m_0 - m_i| < \epsilon \tag{6.23}$$

and

$$|\bar{x}_i^\alpha - \tilde{x}_i^\alpha| \leq C\epsilon_1|m_i - m_0|,$$

which implies

$$|\bar{x}_i^\alpha| \leq C\epsilon_1|m_i - m_0| + |\tilde{x}_i^\alpha|$$

for $\alpha = u, s$ and $i = 1, 2$. So we have

$$|\bar{x}_i^u| \leq C\epsilon_1|m_i - m_0| + |\tilde{x}_i^u|$$
$$\leq C\rho^{-1}\epsilon_1\epsilon + (1 - \eta)^{-1}\rho^{-3}\epsilon < \rho^{-2}\epsilon \tag{6.24}$$

provided that ϵ_1 and η also satisfy $\epsilon_1 < \frac{\rho-1}{C(\rho+1)\rho^2}$ and $\eta < \frac{\rho-1}{2\rho}$. Clearly, we have

$$|\tilde{x}_i^u| \leq (1 - \eta)^{-1}|x_i^u| < \rho^{-1}\epsilon.$$

Note that

$$|\tilde{x}_i^s| \leq |g(m_i, 0)| + \rho\mu|\tilde{x}_i^u| \leq C\delta + \mu\epsilon.$$

Hence,

$$|\bar{x}_i^s| \leq C\epsilon_1|m_i - m_0| + |\tilde{x}_i^s| \leq C\epsilon_1\epsilon + C\delta + \mu\epsilon < \epsilon \tag{6.25}$$

provided that ϵ_1 and δ are sufficiently small and $\mu \in (0, 1)$. By the uniqueness in the proof of the existence of f, (6.23), (6.24), (6.25) give the result.

Finally, we prove that f is Lipschitz. To do that, choose $(\bar{x}_i^u, \bar{x}_i^c) \in \overline{X_{m_0}^u(\rho^{-2}\epsilon)} \times \overline{X_{m_0}^c(\epsilon)}, i = 1, 2$. Write

$$m_i + x_i^u + x_i^s = m_0 + \bar{x}_i^u + \bar{x}_i^s + \bar{x}_i^c,$$

where $\bar{x}_i^s = f(\bar{x}_i^u, \bar{x}_i^c)$. By lemma 4.5, maybe for even smaller ϵ, we can prove that $m_i \in B(m_0, \rho\epsilon)$ and $x_i^u \in X_{m_i}^u(\rho^{-1}\epsilon)$. Because these relationships are the same as those given by (6.3) we may use various estimates from the proof of the previous lemma. In particular, (6.9) gives, for some constant C,

$$|\bar{x}_1^s - \bar{x}_2^s| \leq (C\epsilon + (1 + \eta)^2\mu)(|\bar{x}_1^u - \bar{x}_2^u| + |\bar{x}_1^c - \bar{x}_2^c|)$$
$$\leq \rho\mu(|\bar{x}_1^u - \bar{x}_2^u| + |\bar{x}_1^c - \bar{x}_2^c|), \tag{6.26}$$

which gives the desired Lipschitz property. □

Theorem 6.3. *Let $\lambda_1 \in (\lambda, 1)$ and $\mu \in (0,1)$. Then there exist positive constants $\tilde{\epsilon}^*$, $\epsilon^* = \epsilon^*(\tilde{\epsilon})$, $\delta^* = \delta^*(\epsilon) < \epsilon$ and $\sigma = \sigma(\epsilon, \delta)$ such that if $\tilde{\epsilon} < \tilde{\epsilon}^*$, $\epsilon < \epsilon^*$, $\delta < \delta^*$, and $\tilde{T}$ satisfies $\|\tilde{T} - T^t\|_1 < \sigma$, then $\tilde{T}$ has a Lipschitz center-unstable manifold*

$$\tilde{W}^{cu}(\epsilon) = \Theta(gr(\tilde{h}^{cu})),$$

where $\tilde{h}^{cu} \in \Gamma^{cu}(\epsilon, \mu, \delta, \tilde{\epsilon})$.

The proof of this theorem is based on the following proposition and lemmas. Throughout this section we shall assume that λ_1 and μ satisfy the hypotheses of Theorem 6.3 and $\rho \in (1, 1/\sqrt{\lambda})$ is that used in Lemmas 6.1 and 6.2. The key to finding the center-unstable manifold is to define a "graph" transform whose fixed point gives the center-unstable manifold.

Proposition 6.4. *There exist positive constants $\tilde{\epsilon}^*$, $\epsilon^* = \epsilon^*(\tilde{\epsilon})$, $\delta^* = \delta^*(\epsilon) < \epsilon$ and $\sigma = \sigma(\epsilon, \delta)$ such that if $\tilde{\epsilon} < \tilde{\epsilon}^*$, $\epsilon < \epsilon^*$, $\delta < \delta^*$, and $\tilde{T}$ satisfies $\|\tilde{T} - T^t\|_1 < \sigma$, then for each $h \in \Gamma^{cu}$ there exists a unique $\tilde{h} \in \Gamma^{cu}$ such that*

$$\tilde{T}(\Theta(gr(h))) \cap \Theta(X^u(\epsilon) \oplus X^s(\epsilon)) = \Theta(gr(\tilde{h})) \tag{6.27}$$

To prove this we need some lemmas, the first of which says that $\tilde{T}$ applied to a suitable graph over $X^u(\epsilon)$ still covers $X^u(\epsilon)$.

Lemma 6.5. *There exist positive constants $\tilde{\epsilon}^*$, $\epsilon^* = \epsilon^*(\tilde{\epsilon})$, $\delta^* = \delta^*(\epsilon) < \epsilon$ and $\sigma = \sigma(\epsilon, \delta)$ such that if $\tilde{\epsilon} < \tilde{\epsilon}^*$, $\epsilon < \epsilon^*$, $\delta < \delta^*$, and $\tilde{T}$ satisfies $\|\tilde{T} - T^t\|_1 < \sigma$, then for $h \in \Gamma^{cu}$ and for all $(\bar{m}, \bar{x}^u) \in \overline{X^u(\epsilon)}$ there exists $(m, x^u) \in \overline{X^u(\epsilon)}$ and $\bar{x}^s \in X^s_m(\epsilon)$ satisfying*

$$\tilde{T}(m + x^u + h(m, x^u)) = \bar{m} + \bar{x}^u + \bar{x}^s. \tag{6.28}$$

Proof. The constants $\tilde{\epsilon}^*, \tilde{\epsilon} < \tilde{\epsilon}^*, \epsilon^*(\tilde{\epsilon}) < \tilde{\epsilon}$ and $\delta^*(\epsilon) < \epsilon$ are chosen as in Lemma 6.2 with further restrictions specified later. Because T^t is a diffeomorphism on M, there exists $m_0 \in M$ such that $T^t(m_0) = \bar{m}$. Let $f : \overline{X^u_{m_0}(\rho^{-2}\epsilon)} \times \overline{X^c_{m_0}(\epsilon)} \to X^s_{m_0}$ be that given by Lemma 6.2.

For each $(x^c, x^u) \in \bar{X}^c_{m_0}(\epsilon) \times \bar{X}^u_{m_0}(\rho^{-2}\epsilon)$ write

$$\tilde{T}(m_0 + x^c + x^u + f(x^u, x^c)) = \bar{m} + \tilde{x}^u + \tilde{x}^s + \tilde{x}^c \tag{6.29}$$

for some $\tilde{x}^\alpha \in X^\alpha_{\bar{m}}$, $\alpha = u, s, c$.

To solve (6.28) it is enough to find (x^u, x^c) and $\tilde{x}^s$ such that (6.29) holds for $\tilde{x}^u = \bar{x}^u$ and $\tilde{x}^c = 0$. The linear approximations of (6.29) suggests defining a map $\xi_1 : \overline{X}^c_{m_0}(\epsilon) \times \overline{X}^u_{m_0}(\rho^{-2}\epsilon) \to X^c_{m_0} \times X^u_{m_0}$ by

$$\xi_1(x^c, x^u) \equiv \left(x^c - \left(DT^t(m_0)|_{X^c_{m_0}}\right)^{-1} \tilde{x}^c, x^u - \left(DT^t(m_0)|_{X^u_{m_0}}\right)^{-1} (\tilde{x}^u - \bar{x}^u)\right).$$

Observe that the fixed point of ξ_1 is the solution of (6.29). We claim that ξ_1 is a contraction on $\overline{X^c_{m_0}(\epsilon)} \times \overline{X^u_{m_0}(\rho^{-2}\epsilon)}$. To see this, first take two points $(x^c_i, x^u_i), i = 1, 2$, from this set and observe that with $\tilde{x}^\alpha_i$ given by (6.29)

$$
\begin{aligned}
\tilde{x}^u_1 - \tilde{x}^u_2 &= \Pi^u_{\tilde{m}} \left(\tilde{T}(m_0 + x^c_1 + x^u_1 + f(x^u_1, x^c_1)) - \tilde{T}(m_0 + x^c_2 + x^u_2 + f(x^u_2, x^c_2)) \right) \\
&= \Pi^u_{\tilde{m}} \int_0^1 D\tilde{T}(\ell(\tau)) \left(x^c_1 - x^c_2 + x^u_1 - x^u_2 + f(x^u_1, x^c_1) - f(x^u_2, x^c_2) \right) d\tau
\end{aligned}
$$

where $\ell(\tau) = m_0 + \tau(x^c_1 + x^u_1 + f(x^u_1, x^c_1)) + (1 - \tau)(x^c_2 + x^u_2 + f(x^u_2, x^c_2))$. Then, since $\Pi^u_{\tilde{m}} DT^t(m_0) = 0$ on $X^c_{m_0} \oplus X^s_{m_0}$,

$$
\left| x^u_1 - x^u_2 - \left(DT^t(m_0)|_{X^u_{m_0}} \right)^{-1} (\tilde{x}^u_1 - \tilde{x}^u_2) \right|
$$

$$
\leq \left\| \left(DT^t(m_0)|_{X^u_{m_0}} \right)^{-1} \right\| \cdot |\Pi^u_{\tilde{m}} \int_0^1 \left[DT^t(m_0) - DT^t(\ell(\tau)) + DT^t(\ell(\tau)) - D\tilde{T}(\ell(\tau)) \right]
$$

$$
(x^u_1 - x^u_2 + x^c_1 - x^c_2 + f(x^u_1, x^c_1) - f(x^u_2, x^c_2)) \, d\tau|
$$

$$
\leq \left\| \left(DT^t(m_0))|_{X^u_{m_0}} \right)^{-1} \right\| \|\Pi^u_{\tilde{m}}\| \left[\int_0^1 \|DT^t(m_0) - DT^t(\ell(\tau))\| d\tau + \|T^t - \tilde{T}\|_1 \right]
$$

$$
(1 + \rho\mu) (|x^u_1 - x^u_2| + |x^c_1 - x^c_2|)
$$

$$
\leq C[O(\epsilon) + \sigma] (|x^u_1 - x^u_2| + |x^c_1 - x^c_2|), \tag{6.30}
$$

where the constant C and the term $O(\epsilon)$ do not depend on m_0. A similar estimate holds for $|x^c_1 - x^c_2 - \left(DT^t(m_0)|_{X^c_{m_0}} \right)^{-1} (\tilde{x}^c_1 - \tilde{x}^c_2)|$. Hence, for ϵ^* and σ sufficiently small, ξ_1 is a contraction. We must show that it maps $\overline{X^c_{m_0}(\epsilon)} \times \overline{X^u_{m_0}(\rho^{-2}\epsilon)}$ into itself. When $x^u = x^c = 0$ then for $\alpha = u, c$

$$
\begin{aligned}
\tilde{x}^\alpha &= \Pi^\alpha_{\tilde{m}} \left(\tilde{T}(m_0 + f(0,0)) - \tilde{m} \right) \\
&= \Pi^\alpha_{\tilde{m}} \left(\tilde{T}(m_0 + f(0,0)) - T^t(m_0 + f(0,0)) + T^t(m_0 + f(0,0)) - T^t(m_0) \right),
\end{aligned}
$$

and so for some constant C,

$$
\begin{aligned}
|\tilde{x}^\alpha| &\leq C\|\tilde{T} - T^t\|_0 + |\Pi^\alpha_{\tilde{m}} \int_0^1 \left(DT^t(m_0 + \tau f(0,0)) - DT^t(m_0) \right) f(0,0) d\tau| \\
&< C\sigma + \delta O(\delta).
\end{aligned}
$$

Also, by (H3)

$$
\left| \left(DT^t(m_0)|_{X^u_{m_0}} \right)^{-1} \tilde{x}^u \right| \leq \lambda |\tilde{x}^u| < \lambda \epsilon.
$$

If ξ_1^c and ξ_1^u are the components of ξ_1, then these give

$$|\xi_1^c(0,0)| \le C\sigma + \delta O(\delta) \tag{6.31}$$

and

$$|\xi_1^u(0,0)| \le \lambda\epsilon + C\sigma + \delta O(\delta). \tag{6.32}$$

Note that $\delta < \epsilon$. For $(x^c, x^u) \in \overline{X_{m_0}^c(\epsilon)} \times \overline{X_{m_0}^u(\rho^{-2}\epsilon)}$, (6.30) and the corresponding estimate for ξ_1^c combine with (6.31) and (6.32) to give

$$|\xi_1^u(x^c, x^u)| \le |\xi_1^u(0,0)| + |\xi_1^u(x^c, x^u) - \xi_1^u(0,0)|$$
$$< \lambda\epsilon + C\sigma + \delta O(\delta) + C[O(\epsilon) + \sigma](\rho^{-2}\epsilon + \epsilon) < \rho^{-2}\epsilon$$

and

$$|\xi_1^c(x^c, x^u)| \le |\xi_1^c(0,0)| + |\xi_1^c(x^c, x^u) - \xi_1^c(0,0)|$$
$$< C\sigma + \delta O(\delta) + C[O(\epsilon) + \sigma](\rho^{-2}\epsilon + \epsilon) < \epsilon$$

by choosing ϵ^* and σ sufficiently small, $\delta < \epsilon$ and $\rho \in (1, \frac{1}{\sqrt{\lambda}})$.

Having proved that ξ_1 is a contraction on $\overline{X_{m_0}^c(\epsilon)} \times \overline{X_{m_0}^u(\rho^{-2}\epsilon)}$ it follows that ξ_1 has a unique fixed point (x_0^c, x_0^u) in that set. This implies that the corresponding points given by (6.29) satisfy

$$\tilde{x}^c = 0 \text{ and } \tilde{x}^u = \bar{x}^u$$

and so

$$\tilde{T}(m_0 + x_0^u + x_0^c + f(x_0^c, x_0^u)) = \bar{m} + \bar{x}^u + \tilde{x}^s.$$

This completes the proof of the lemma. $\qquad\square$

Notice that the uniqueness of the fixed point of ξ_1 in $\overline{X_{m_0}^c(\epsilon)} \times \overline{X_{m_0}^u(\rho^{-2}\epsilon)}$ does not imply the uniqueness of the point $\bar{x}^s$ in (6.28). But it is a consequence of the following lemma.

Let

$$\tilde{T}(m_0 + x_0^u + x_0^s) = m_1 + x_1^u + x_1^s \tag{6.33}$$

where $x_0^s = h(m_0, x_0^u)$ and m_0, x_0^u, x_1^s are determined by (m_1, x_1^u) as in Lemma 6.5.

Consider a perturbation of $m_1 + x_1^u + x_1^s$

$$m_1 + x_1^u + x_1^s + \tilde{x}^u + \tilde{x}^s + \tilde{x}^c$$

where $\tilde{x}^\alpha \in \overline{X_{m_1}^\alpha(\bar{\epsilon})}$, for $\alpha = u, s, c$.

Lemma 6.6. *There exist positive constants* $\tilde{\epsilon}^*$, $\epsilon^* = \epsilon^*(\tilde{\epsilon})$, $\delta^* = \delta^*(\epsilon) < \epsilon$ *and* $\sigma = \sigma(\epsilon, \delta)$ *such that for* $\tilde{\epsilon} < \tilde{\epsilon}^*$, $\epsilon < \epsilon^*$, $\delta < \delta^*$, *and* $\tilde{T}$ *satisfying* $\|\tilde{T} - T^t\|_1 < \sigma$, *if*

$$m_1 + x_1^u + x_1^s + \tilde{x}^u + \tilde{x}^s + \tilde{x}^c \in \Theta(\overline{X^u(\epsilon)} \oplus \overline{X^s(\epsilon)})$$

and if there exists $(m, x^u) \in \overline{X^u(\epsilon)}$ *such that*

$$\tilde{T}(m + x^u + h(m, x^u)) = m_1 + x_1^u + x_1^s + \tilde{x}^u + \tilde{x}^s + \tilde{x}^c$$

then

$$|\tilde{x}^s| \le \mu(|\tilde{x}^u| + |\tilde{x}^c|).$$

Proof. From the assumption of the lemma we may write

$$m_1 + x_1^u + x_1^s + \tilde{x}^u + \tilde{x}^s + \tilde{x}^c = m_2 + x_2^u + x_2^s \tag{6.34}$$

where $x_2^\alpha \in X^\alpha(\epsilon), \alpha = u, s.$. Thus,

$$\tilde{T}(m + x^u + x^s) = m_2 + x_2^u + x_2^s \tag{6.35}$$

where $x^s = h(m, x^u)$.

We prove this lemma by contradiction. Suppose that

$$|\tilde{x}^s| > \mu(|\tilde{x}^u| + |\tilde{x}^c|). \tag{6.36}$$

First, from (6.34)

$$|m_2 - m_1| \le C\tilde{\epsilon}.$$

Choose $\tilde{\epsilon}^*$ sufficiently small so that m_1 and m_2 are in η-neighborhoods of each other, where η satisfies the conditions in the proof of Lemma 6.2. Then, from (4.17), using (6.34) and the point $m_1 + x_1^u + x_1^s$,

$$\left| x_1^s + \tilde{x}^s - \left(\Pi_{m_2}^s |_{X_{m_1}^s} \right)^{-1} x_2^s \right| \le C\epsilon_1 |m_2 - m_1| \le C\epsilon_1\tilde{\epsilon},$$

and so

$$|\tilde{x}^s| \le C\epsilon + C\epsilon_1\tilde{\epsilon}. \tag{6.37}$$

Here we used the fact that $|x_i^\alpha| < \epsilon$ for $\alpha = u, s$ and $i = 1, 2$.

Now (6.36) gives the estimates $|\tilde{x}^\alpha| \le \frac{C}{\mu}(\epsilon + \epsilon_1\tilde{\epsilon})$ for $\alpha = u, c$.

Since T^t is a diffeomorphism on M we have, for some constant C, using the above estimates

$$|m - m_0| \le C|T^t(m) - T^t(m_0)| \le C\left(|\tilde{T}(m) - \tilde{T}(m_0)| + 2\|T^t - \tilde{T}\|_0 \right)$$

$$\le C\left(|\tilde{T}(m) - \tilde{T}(m + x^u + x^s)| + |\tilde{T}(m_0) - \tilde{T}(m_0 + x_0^u + x_0^s)| + \right.$$

$$\left. |\tilde{T}(m + x^u + x^s) - \tilde{T}(m_0 + x_0^u + x_0^s)| + 2\sigma \right)$$

$$\le \frac{C}{\mu}(\epsilon + \sigma + \epsilon_1\tilde{\epsilon}). \tag{6.38}$$

Thus, we may choose $\tilde{\epsilon}^*$, ϵ^*, and σ so small that m is in an η-neighborhood of m_0. Applying Lemma 4.5 to m_0 and the point

$$m + x^u + x^s = m_0 + x_0^u + x_0^s + \bar{x}^u + \bar{x}^s + \bar{x}^c$$

we have

$$|\bar{x}^\alpha| < \frac{C}{\mu}(\epsilon + \sigma + \epsilon_1 \tilde{\epsilon}) < \tilde{\epsilon}$$

by choosing $\epsilon_1 < \mu/2C$, $\epsilon^* < \mu\tilde{\epsilon}/4C$, and $\sigma^* < \mu\tilde{\epsilon}/4C$.

Now, because $h \in \Gamma^{cu}$

$$|\bar{x}^s| \leq \mu(|\bar{x}^u| + |\bar{x}^c|).$$

But then the moving cone "invariance," given by Lemma 5.5, gives

$$|\bar{x}^s| \leq \mu(|\bar{x}^u| + |\bar{x}^c|),$$

which contradicts (6.36). The proof is complete. $\square$

An immediate consequence of this lemma is the following

Corollary 6.7. *There exist positive constants* $\tilde{\epsilon}^*$, $\epsilon^* = \epsilon^*(\tilde{\epsilon})$, $\delta^* = \delta^*(\epsilon) < \epsilon$ *and* $\sigma = \sigma(\epsilon, \delta)$ *such that if* $\tilde{\epsilon} < \tilde{\epsilon}^*$, $\epsilon < \epsilon^*$, $\delta < \delta^*$, *and* $\tilde{T}$ *satisfies* $\|\tilde{T} - T^t\|_1 < \sigma$, *then* $\bar{x}^s$ *in (6.28) is unique.*

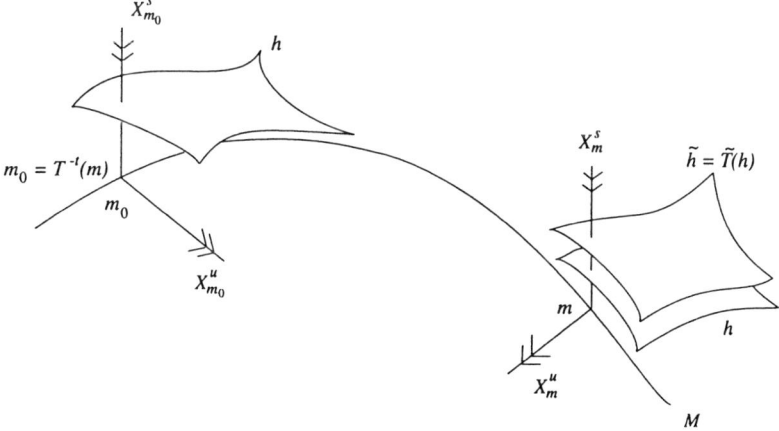

Figure 5. The Graph Transform

The uniqueness of $\bar{x}^s$ allows us to define a map

$$\tilde{h}(\bar{m}, \bar{x}^u) = \bar{x}^s$$

and so we have a graph transform but it remains to show $\tilde{h} \in \Gamma^{cu}$.

Proof of Proposition 6.4. It is obvious that

$$\tilde{T}(\Theta(gr(h))) \supset \Theta(gr(\tilde{h})).$$

Corollary 6.7 also implies the uniqueness of $\tilde{h}$ and that (6.27) holds. By using Lemma 5.2 and Lemma 6.6, we find $\tilde{h} \in \Gamma^{cu}$. This completes the proof. $\qquad\square$

From Proposition 6.4, we may define a graph transform $\mathcal{F}^{cu} : \Gamma^{cu} \to \Gamma^{cu}$ by

$$\tilde{h}(\bar{m}, \bar{x}^u) = (\mathcal{F}^{cu}h)(\bar{m}, \bar{x}^u) = \bar{x}^s$$

with $\bar{x}^s$ given by (6.28) in Lemma 6.5. We shall see that $\mathcal{F}^{cu}$ is a contraction.

Before we prove this, we state a lemma which will be used elsewhere. Let $(m_0, x_0^u) \in X^u(\epsilon)$ and $x_1^s, x_2^s \in X_{m_0}^s(\epsilon)$. Consider points $(\tilde{m}_1, \tilde{x}_1^u + \tilde{x}_1^s) \in X^u(\epsilon) \oplus X^s(\epsilon)$ and $(\tilde{m}_2, \tilde{x}_2^u) \in X^u(\epsilon)$ satisfying

$$\tilde{T}(\tilde{m}_1 + \tilde{x}_1^u + \tilde{x}_1^s) = m_0 + x_0^u + x_1^s \tag{6.39}$$

and

$$\tilde{T}(\tilde{m}_2 + \tilde{x}_2^u + h(\tilde{m}_2, \tilde{x}_2^u)) = m_0 + x_0^u + x_2^s, \tag{6.40}$$

where $h \in \Gamma^{cu}$. Let $\lambda_1 \in (\lambda, 1)$.

Lemma 6.8. *There exist positive constants $\tilde{\epsilon}^*$, $\epsilon^* = \epsilon^*(\tilde{\epsilon})$, $\delta^* = \delta^*(\epsilon) < \epsilon$ and $\sigma = \sigma(\epsilon, \delta)$ such that if $\tilde{\epsilon} < \tilde{\epsilon}^*$, $\epsilon < \epsilon^*$, $\delta < \delta^*$, and $\tilde{T}$ satisfies $\|\tilde{T} - T^t\|_1 < \sigma$, then*

$$|x_1^s - x_2^s| \leq \lambda_1 |\tilde{x}_1^u - h(\tilde{m}_1, \tilde{x}_1^u)|.$$

Proof. Let $\tilde{x}_2^s = h(\tilde{m}_2, \tilde{x}_2^u)$. Using an argument similar to that giving (6.38) we find $|\tilde{m}_1 - \tilde{m}_2| \leq C(\epsilon + \sigma + \epsilon_1 \tilde{\epsilon})$ for some $C > 0$. From this we may use the representation

$$\tilde{m}_2 + \tilde{x}_2^u + \tilde{x}_2^s = \tilde{m}_1 + \tilde{x}_1^u + \tilde{x}_1^s + \bar{x}^u + \bar{x}^s + \bar{x}^c,$$

where $\bar{x}^\alpha \in X_{\tilde{m}_1}^\alpha$. Clearly $|\bar{x}^\alpha| \leq C(\sigma + \epsilon + \epsilon_1 \tilde{\epsilon})$ for $\alpha = u, s, c$.

We wish to estimate $|x_1^s - x_2^s|$. First we write $\ell(\tau) = \tilde{m}_1 + \tilde{x}_1^u + \tilde{x}_1^s + \tau(\bar{x}^u + \bar{x}^s + \bar{x}^c)$ so that

$$x_1^s - x_2^s = \tilde{T}(\ell(0)) - \tilde{T}(\ell(1)) = \int_0^1 D\tilde{T}(\ell(\tau))(\bar{x}^u + \bar{x}^s + \bar{x}^c)d\tau. \tag{6.41}$$

To estimate the right hand side, note that

$$0 = \Pi_{m_0}^u \int_0^1 D\tilde{T}(\ell(\tau))(\bar{x}^u + \bar{x}^s + \bar{x}^c)d\tau$$

$$= \Pi_{m_0}^u \int_0^1 \left[D\tilde{T}(\ell(\tau)) - DT^t(\ell(\tau)) + DT^t(\ell(\tau)) - DT^t(\tilde{m}_1) \right] (\bar{x}^u + \bar{x}^s + \bar{x}^c)d\tau$$

$$+ \Pi_{m_0}^u DT^t(\tilde{m}_1)(\bar{x}^u + \bar{x}^s + \bar{x}^c).$$

So

$$|\Pi^u_{m_0} DT^t(\tilde{m}_1)\bar{x}^u|$$
$$\leq |\left(\Pi^u_{m_0} - \Pi^u_{\hat{m}_2}\right) DT^t(\tilde{m}_1)(\bar{x}^s + \bar{x}^c)|$$
$$+ \|\Pi^u_{m_0}\|[\sigma + O(\epsilon + \tilde{\epsilon} + \sigma)](|\bar{x}^u| + |\bar{x}^s| + |\bar{x}^c|) \tag{6.42}$$

where $\hat{m}_2 \equiv T^t(\tilde{m}_1)$.

Since

$$|m_0 - \hat{m}_2| = |\tilde{T}(\tilde{m}_1 + \tilde{x}^u_1 + \tilde{x}^s_1) - x^u_0 - x^s_1 - T^t(\tilde{m}_1)|$$
$$\leq \|\tilde{T} - T^t\|_0 + |T^t(\tilde{m}_1 + \tilde{x}^u_1 + \tilde{x}^s_1) - T^t(\tilde{m}_1)| + |x^u_0| + |x^s_1|$$
$$\leq \sigma + C\epsilon$$

for some $C > 0$,

$$\|\Pi^u_{m_0} - \Pi^u_{\hat{m}_2}\| \to 0$$

and

$$\|\left(\Pi^u_{m_0}|_{X^u_{\hat{m}_2}}\right)^{-1}\| \to 1$$

as $\sigma + \epsilon \to 0$. Therefore, from (6.42) and (H3)

$$|\bar{x}^u| \leq \lambda |DT^t(\tilde{m}_1)\bar{x}^u|$$
$$\leq \lambda \|\left(\Pi^u_{m_0}|_{X^u_{\hat{m}_2}}\right)^{-1}\| \|\Pi^u_{m_0} DT^t(\tilde{m}_1)\bar{x}^u|$$
$$= O(\sigma + \epsilon + \tilde{\epsilon})(|\bar{x}^u| + |\bar{x}^s| + |\bar{x}^c|) \tag{6.43}$$

as $\rho + \epsilon \to 0$. A similar estimate holds for $|\bar{x}^c|$ and consequently,

$$|\bar{x}^u| + |\bar{x}^c| = O(\sigma + \epsilon + \tilde{\epsilon})|\bar{x}^s|. \tag{6.44}$$

Write

$$\tilde{m}_2 + \tilde{x}^u_2 + \tilde{x}^s_2 = \tilde{m}_1 + \tilde{x}^u_1 + h(\tilde{m}_1, \tilde{x}^u_1) + \bar{x}^u + [\tilde{x}^s_1 + \bar{x}^s - h(\tilde{m}_1, \tilde{x}^u_1)] + \bar{x}^c$$

By choosing ϵ_1, ϵ^* and σ sufficiently small, we obtain $|\bar{x}^u| < \tilde{\epsilon}$, $|\tilde{x}^s_1 + \bar{x}^s - h(\tilde{m}_1, \tilde{x}^u_1)| < \tilde{\epsilon}$ and $|\bar{x}^c| < \tilde{\epsilon}$.

Thus, using the fact that $h \in \Gamma^{cu}$ we have

$$\mu(|\bar{x}^u| + |\bar{x}^c|) \geq |\tilde{x}^s_1 + \bar{x}^s - h(\tilde{m}_1, \tilde{x}^u_1)|$$
$$\geq |\bar{x}^s| - |\tilde{x}^s_1 - h(\tilde{m}_1, \tilde{x}^u_1)|.$$

This with (6.44) gives

$$|\bar{x}^s| \leq (1 + O(\sigma + \epsilon + \tilde{\epsilon}))|\tilde{x}^s_1 - h(\tilde{m}_1, \tilde{x}^u_1)|.$$

(6.41) also yields, in a manner similar to that which produced (6.43),

$$|x_1^s - x_2^s| < (\lambda + O(\sigma + \epsilon + \tilde{\epsilon}))|\bar{x}^s|.$$

Consequently,

$$|x_1^s - x_2^s| < (\lambda + O(\sigma + \epsilon + \tilde{\epsilon}))(1 + C(\sigma + \epsilon + \tilde{\epsilon}))|\tilde{x}_1^s - h(\tilde{m}_1, \tilde{x}_1^u)|$$
$$\leq \lambda_1 |\tilde{x}_1^s - h(\tilde{m}_1, \tilde{x}_1^u)|$$

for ϵ^* and σ sufficiently small. This completes the proof. $\qquad\square$

Proposition 6.9. *There exist positive constants $\tilde{\epsilon}^*$, $\epsilon^* = \epsilon^*(\tilde{\epsilon})$, $\delta^* = \delta^*(\epsilon) < \epsilon$ and $\sigma = \sigma(\epsilon, \delta)$ such that if $\tilde{\epsilon} < \tilde{\epsilon}^*$, $\epsilon < \epsilon^*$, $\delta < \delta^*$, and $\tilde{T}$ satisfies $\|\tilde{T} - T^t\|_1 < \sigma$, then $\mathcal{F}^{cu}$ is a contraction on Γ^{cu}.*

Proof. With $h_1, h_2 \in \Gamma^{cu}$ and $(m_0, x_0^u) \in X^u(\epsilon)$, for $i = 1, 2$ let

$$x_i^s = (\mathcal{F}^{cu} h_i)(m_0, x_0^u).$$

From the definition of $\mathcal{F}^{cu}$, there exists $(\tilde{m}_i, \tilde{x}_i^u) \in X^u(\epsilon)$ such that

$$\tilde{T}(\tilde{m}_1 + \tilde{x}_1^u + h_1(\tilde{m}_1, \tilde{x}_1^u)) = m_0 + x_0^u + x_1^s$$

and

$$\tilde{T}(\tilde{m}_2 + \tilde{x}_2^u + h_2(\tilde{m}_2, \tilde{x}_2^u)) = m_0 + x_0^u + x_2^s.$$

Regarding $h_1(\tilde{m}_1, \tilde{x}_1^u)$ as $\tilde{x}_1^s$ in (6.39) and using Lemma 6.8, we obtain

$$|x_1^s - x_2^s| \leq \lambda_1 \|h_1 - h_2\|.$$

Since $\lambda_1 < 1$, $\mathcal{F}^{cu}$ is a contraction. This completes the proof. $\qquad\square$

Proof Theorem 6.3. From Proposition 6.9, $\mathcal{F}^{cu}$ is a contraction, hence has a fixed point in Γ^{cu} which we denote by $\tilde{h}^{cu}$. Clearly,

$$\tilde{T}(\Theta(gr(\tilde{h}^{cu}))) \cap K_f(\epsilon, \mu, \delta) = \Theta(gr(\tilde{h}^{cu})).$$

Thus $\tilde{W}^{cu}(\epsilon) \equiv \Theta(gr(\tilde{h}^{cu}))$ is a local invariant center-unstable manifold for $\tilde{T}$. This completes the proof. $\qquad\square$

Now we consider another characterization of $\tilde{W}^{cu}(\epsilon)$. Define by induction a sequence of sets $\mathcal{A}_k, k = 1, 2, \cdots$ by

$$\mathcal{A}_k = \tilde{T}(\mathcal{A}_{k-1}) \cap \Theta(X^u(\epsilon) \oplus X^s(\epsilon))$$

for $k \geq 1$ and $\mathcal{A}_0 = \Theta(X^u(\epsilon) \oplus X^s(\epsilon))$. It is clear that we may also write

$$\mathcal{A}_k = \{x_0 \in \Theta(X^u(\epsilon) \oplus X^s(\epsilon)) : \text{there exists } x_k \in \Theta(X^u(\epsilon) \oplus X^s(\epsilon))$$
$$\text{such that } x_l = \tilde{T}^{(k-l)} x_k \in \Theta(X^u(\epsilon) \oplus X^s(\epsilon)) \text{ for } 0 \leq l \leq k\}.$$

Proposition 6.10.

$$\tilde{W}^{cu}(\epsilon) = \cap_{k=1}^{\infty} \mathcal{A}_k$$

Proof. We first show that $\tilde{W}^{cu}(\epsilon) \subset \cap_{k=1}^{\infty} \mathcal{A}_k$. For each $m_0 + x_0^u + \tilde{h}^{cu}(m_0, x_0^u) \in \tilde{W}^{cu}(\epsilon)$, since $\tilde{h}^{cu}$ is the fixed point of $\mathcal{F}^{cu}$, there exists $(m_1, x_1^u) \in X^u(\epsilon)$ such that

$$m_0 + x_0^u + \tilde{h}^{cu}(m_0, x_0^u) = \tilde{T}(m_1 + x_1^u + \tilde{h}^{cu}(m_1, x_1^u)),$$

which yields that $m_0 + x_0^u + \tilde{h}^{cu}(m_0, x_0^u) \in \mathcal{A}_k$ for all $k \geq 1$. Next we show $\cap_{k=1}^{\infty} \mathcal{A}_k \subset \tilde{W}^{cu}(\epsilon)$. Let $m + x^u + x^s \in \cap_{k=1}^{\infty} \mathcal{A}_k$. We claim that $x^s = \tilde{h}^{cu}(m, x^u)$. For each $k \geq 1$, since $m + x^u + x^s \in \mathcal{A}_k$, there exists $m_0 + x_0^u + x_0^s \in \Theta(X^u(\epsilon) \oplus X^s(\epsilon))$ such that

$$\tilde{T}^k(m_0 + x_0^u + x_0^s) = m + x^u + x^s$$

and

$$\tilde{T}^i(m_0 + x_0^u + x_0^s) \in \Theta(X^u(\epsilon) \oplus X^s(\epsilon)), \ 1 \leq i \leq k.$$

For $i = 1, 2, \cdots, k$, let

$$m_i + x_i^u + x_i^s = \tilde{T}^i(m_0 + x_0^u + x_0^s). \tag{6.45}$$

Note that $m_k + x_k^u + x_k^s = m + x^u + x^s$. Since $\tilde{h}^{cu}$ is the fixed point of $\mathcal{F}^{cu}$, there exists $(\bar{m}_0, \bar{x}_0^u) \in X^u(\epsilon)$ such that

$$\tilde{T}(\bar{m}_0 + \bar{x}_0^u + \tilde{h}^{cu}(\bar{m}_0, \bar{x}_0^u)) = m_1 + x_1^u + \tilde{h}^{cu}(m_1, x_1^u).$$

Let $i = 1$ in (6.45)

$$\tilde{T}(m_0 + x_0^u + x_0^s) = m_1 + x_1^u + x_1^s.$$

From Lemma 6.8, we obtain

$$|x_1^s - \tilde{h}^{cu}(m_1, x_1^u)| \leq \lambda_1 |x_0^s - \tilde{h}^{cu}(m_0, x_0^u)|.$$

Repeating this procedure gives

$$|x^s - \tilde{h}^{cu}(m, x^u)| = |x_k^s - \tilde{h}^{cu}(m_k, x_k^u)| \leq \lambda_1^k |x_0^s - \tilde{h}^{cu}(m_0, x_0^u)| \leq 2\epsilon\lambda_1^k, \tag{6.46}$$

which yields $x^s = \tilde{h}^{cu}(m, x^u)$ by letting $k \to \infty$. The proof is complete. $\square$

From Theorem 6.3 there exists a center-unstable manifold for the time-t map T^t given by

$$W^{cu}(\epsilon) = \Theta(gr(h^{cu}))$$

where $h^{cu} \in \Gamma^{cu}$. The next result states that the perturbed center-unstable manifold $\tilde{W}^{cu}(\epsilon)$ is close to the unperturbed one, $W^{cu}(\epsilon)$.

Proposition 6.11. *There exists ϵ^* such that for $\epsilon < \epsilon^*$,*

$$\|\tilde{h}^{cu} - h^{cu}\| \to 0 \quad as \quad \|\tilde{T} - T^t\|_0 \to 0.$$

Proof. We denote the ϵ^* in Theorem 6.3 temporarily by ϵ_1^*. Let $\epsilon_2 < \min\{\epsilon_1^*, \frac{1}{2C}\}$ be fixed, where C is the constant in Lemma 4.5. From Lemma 4.4, there is a β^* such that (4.12) holds with ϵ_1 instead of ϵ. Let η and r satisfy (4.14). By Lemma 4.3, we may choose ϵ^* sufficiently small such that for $\epsilon < \epsilon^*$, $(m_1, x_1^u + x_1^s) \in X^u(\epsilon) \oplus X^s(\epsilon)$, and for $x \in X$ if $|m_1 + x_1^u + x_1^s - x| < \epsilon$, then $x = m_2 + x_2^u + x_2^s$, where $x_2^\alpha \in X^\alpha(\epsilon_2)$ and m_2 is in an η-neighborhood of m_1.

Note that Proposition 6.10 implies that $\tilde{W}^{cu}(\epsilon) \subset \tilde{W}^{cu}(\epsilon_2)$ for all $\epsilon < \epsilon_2 < \epsilon_1^*$. In other words, if we denote $\tilde{h}^{cu}$ by $\tilde{h}_\epsilon^{cu}$, then we have $\tilde{h}_\epsilon^{cu} = \tilde{h}_{\epsilon_2}^{cu}$ on $X^u(\epsilon)$.

Let $\mathcal{F}^{cu}$ denote the graph transform for the time-t map T^t and $\tilde{\mathcal{F}}^{cu}$ that for $\tilde{T}$. From Proposition 6.9, $\mathcal{F}^{cu}$ is a contraction in $\Gamma^{cu}(\epsilon_2, \mu, \delta, \tilde{\epsilon})$ and h^{cu} is its fixed point. Thus, for any $\tilde{h}^{cu} \in \Gamma^{cu}$ and for small $\mathcal{E} > 0$, there is a positive integer k such that

$$\left\|\left(\mathcal{F}^{cu}\right)^k(\tilde{h}^{cu}) - h^{cu}\right\| \leq \mathcal{E}. \tag{6.47}$$

For such fixed k, it is easy to see that there exists $\beta > 0$ such that if $\|\tilde{T} - T^t\|_0 \leq \beta$ then for $i = 1, 2, \cdots, k$

$$\|\tilde{T}^i - (T^t)^i\|_0 \leq \mathcal{E}. \tag{6.48}$$

Let ϵ^* also satisfy $\epsilon^* < \epsilon_2$. Take $\tilde{h}^{cu}$ to be the fixed point of $\tilde{\mathcal{F}}^{cu}$ in $\Gamma^{cu}(\epsilon, \mu, \delta, \tilde{\epsilon})$. For each $(m_0, x_0^u) \in X^u(\epsilon)$, let $x_0^s = \left(\tilde{\mathcal{F}}^{cu}\right)^k \tilde{h}^{cu}(m_0, x_0^u)$ and $\hat{x}_0^s = \left(\mathcal{F}^{cu}\right)^k \tilde{h}^{cu}(m_0, x_0^u)$. We want to estimate $|x_0^s - \hat{x}_0^s|$. From the definition of $\tilde{\mathcal{F}}^{cu}$ there exists $(m_1, x_1^u) \in X^u(\epsilon)$ such that

$$\tilde{T}^k(m_1 + x_1^u + \tilde{h}^{cu}(m_1, x_1^u)) = m_0 + x_0^u + x_0^s. \tag{6.49}$$

Using (6.48) we may write

$$(T^t)^k(m_1 + x_1^u + \tilde{h}^{cu}(m_1, x_1^u)) = \bar{m}_0 + \bar{x}_0^u + \bar{x}_0^s, \tag{6.50}$$

where $\bar{x}_0^\alpha \in X_{\bar{m}_0}^\alpha(\epsilon_2)$. From (6.27) and the definition of $\mathcal{F}^{cu}$ it follows that

$$\bar{x}_0^s = \left(\mathcal{F}^{cu}\right)^k \tilde{h}^{cu}(\bar{m}_0, \bar{x}_0^u). \tag{6.51}$$

Write

$$\bar{m}_0 + \bar{x}_0^u + \bar{x}_0^s = m_0 + x_0^u + x^u + x_0^s + x^s + x^c. \tag{6.52}$$

Then we have

$$x^\alpha = \Pi_{m_0}^\alpha(\bar{m}_0 + \bar{x}_0^u + \bar{x}_0^s - (m_0 + x_0^u + x_0^s)).$$

By (6.48) we obtain

$$|x^\alpha| \le C\mathcal{E} \le \rho^{-3}\epsilon_2 - \epsilon \tag{6.53}$$

for $\alpha = c, s, u$. Thus, we have $x_0^s + x^s = f(x_0^u + x^u, x^c)$, where f is the representation of $\left(\mathcal{F}^{cu}\right)^k \tilde{h}^{cu}$ at m_0. Also $\hat{x}_0^s = f(x_0^u, 0)$, and so by Lemma 6.2

$$|\hat{x}_0^s - (x^s + x_0^s)| \le C\mu(|x^u| + |x^c|).$$

Hence,

$$|\hat{x}_0^s - x_0^s| \le C\mathcal{E}. \tag{6.54}$$

Using (6.47) and (6.54) we obtain

$$|\tilde{h}^{cu}(m_0, x_0^u) - h^{cu}(m_0, x_0^u)|$$
$$\le |\left(\tilde{\mathcal{F}}^{cu}\right)^k (\tilde{h}^{cu})(m_0, x_0^u) - \left(\mathcal{F}^{cu}\right)^k (\tilde{h}^{cu})(m_0, x_0^u)| + |\left(\mathcal{F}^{cu}\right)^k (\tilde{h}^{cu})(m_0, x_0^u) - h^{cu}(m_0, x_0^u)|$$
$$\le C\mathcal{E},$$

which completes the proof. $\qquad\square$

Proposition 6.12. *There exists ϵ^* and σ such that if $\epsilon < \epsilon^*$ and $\tilde{T}$ satisfies $\|\tilde{T} - T^t\|_1 < \sigma$, then*

$$\tilde{T} : \tilde{W}^{cu}(\epsilon) \cap \tilde{T}^{-1}(\tilde{W}^{cu}(\epsilon)) \to \tilde{W}^{cu}(\epsilon)$$

is a homeomorphism.

Proof. From Proposition 6.4, $\tilde{T}$ is onto. We need to show that it is one-to-one and its inverse is continuous. Consider $m_0 + x_0^u + \tilde{h}^{cu}(m_0, x_0^u), m + x^u + \tilde{h}^{cu}(m, x^u) \in \tilde{W}^{cu}(\epsilon) \cap \tilde{T}^{-1}(\tilde{W}^{cu}(\epsilon))$. We may write

$$\tilde{T}(m_0 + x_0^u + \tilde{h}^{cu}(m_0, x_0^u)) = \bar{m}_0 + \bar{x}_0^u + \tilde{h}^{cu}(\bar{m}_0, \bar{x}_0^u) \tag{6.55}$$

and

$$\tilde{T}(m + x^u + \tilde{h}^{cu}(m, x^u)) = \bar{m} + \bar{x}^u + \tilde{h}^{cu}(\bar{m}, \bar{x}^u), \tag{6.56}$$

where $(\bar{m}_0, \bar{x}_0^u), (\bar{m}, \bar{x}^u) \in X^u(\epsilon)$. Let $\zeta > 0$ be given and small. Let

$$\Delta = \bar{m}_0 + \bar{x}_0^u + \tilde{h}^{cu}(\bar{m}_0, \bar{x}_0^u) - (\bar{m} + \bar{x}^u + \tilde{h}^{cu}(\bar{m}, \bar{x}^u))$$

and assume that $|\Delta| \le \zeta$. From (6.38) it follows that

$$|m - m_0| \le C(\epsilon + \sigma + \zeta). \tag{6.57}$$

Thus, we may choose ϵ^*, σ and ζ sufficiently small such that m is in an η-neighborhood of m_0. Write

$$m + x^u + \tilde{h}^{cu}(m, x^u) = m_0 + x_0^u + \tilde{h}^{cu}(m_0, x_0^u) + \tilde{x}^u + \tilde{x}^s + \tilde{x}^c, \tag{6.58}$$

where $x^\alpha \in X^\alpha_{m_0}$. Then applying Lemma 4.5 to m_0 and (6.58), we have

$$|\tilde{x}^\alpha| \leq C(\epsilon + \sigma + \zeta) < \tilde{\epsilon},$$

provided ϵ^*, σ and ζ are sufficiently small. Now, because $\tilde{h}^{cu} \in \Gamma^{cu}(\epsilon, \mu, \delta, \tilde{\epsilon})$

$$|\tilde{x}^s| \leq \mu(|\tilde{x}^u| + |\tilde{x}^c|). \tag{6.59}$$

Using (6.58) and applying the Taylor expansion to (6.56) at $m_0 + x_0^u + \tilde{h}^{cu}(m_0, x_0^u)$, we obtain

$$D\tilde{T}(m_0 + x_0^u + \tilde{h}^{cu}(m_0, x_0^u))(\tilde{x}^u + \tilde{x}^s + \tilde{x}^c) = o(|\tilde{x}^u| + |\tilde{x}^s| + |\tilde{x}^c|) + \Delta. \tag{6.60}$$

Note that
$$\|D\tilde{T}(m_0 + x_0^u + \tilde{h}^{cu}(m_0, x_0^u)) - DT^t(m_0)\| \leq C(\sigma + \epsilon)$$

for some constant C. Hence, from (6.59) and (6.60), we obtain for $\alpha = u, c$,

$$|DT^t(m_0)\tilde{x}^\alpha| \leq O(\sigma + \epsilon + \zeta)(|\tilde{x}^u| + |\tilde{x}^c|) + C\zeta.$$

From (H3), we know that both $\Pi^u_{T^t(m_0)}DT^t(m_0)|_{X^u_{m_0}}$ and $\Pi^c_{T^t(m_0)}DT^t(m_0)|_{X^c_{m_0}}$ have bounded inverses, thus

$$|\tilde{x}^u| + |\tilde{x}^c| \leq O(\sigma + \epsilon + \zeta)(|\tilde{x}^u| + |\tilde{x}^c|) + C\zeta.$$

Choosing ϵ^*, σ and ζ sufficiently small, we obtain

$$|\tilde{x}^u| + |\tilde{x}^c| \leq C\zeta,$$

which with (6.58) implies that $\tilde{T} : \tilde{W}^{cu}(\epsilon) \cap \tilde{T}^{-1}(\tilde{W}^{cu}(\epsilon)) \to \tilde{W}^{cu}(\epsilon)$ is one-to-one and has a continuous inverse. This completes the proof. □

§7. Center-stable manifold.

In section 6, we proved the existence of a Lipschitz center-unstable manifold. In addition to the difficulties addressed in Section 6 we have a further difficulty in constructing the center-stable manifold.

For flows which exist in backward as well as forward time, such as for finite dimensional dynamical systems or hyperbolic PDEs, one may now obtain the center-stable manifold simply by reversing time (see [F] and [HPS]). However, this trick does not work for infinite dimensional dynamical systems in general, like, for example, those generated by parabolic PDE's, since backward solutions may not exist and the corresponding solution map is not invertible. In order to overcome this difficulty, a new technique is needed to establish the existence of the center-stable manifold. We show that for any Lipschitz graph over the stable bundle and any unstable fiber, there is a unique point on the fiber which is mapped into this graph. We then show that the collection of such points is a Lipschitz graph.

As in Section 6, we shall first introduce a space of Lipschitz sections of the bundle $X^u(\epsilon) \oplus X^s(\epsilon)$ over $X^s(\epsilon)$, then construct a graph transform defined on this space, which is induced by $\tilde{T}$. In fact this involves a type of inverse of $\tilde{T}$ even though $\tilde{T}^{-1}$ is not everywhere defined. We then show that this graph transform is a contraction and its fixed point gives the center-stable manifold.

Let h be a section of the bundle $X^u(\epsilon) \oplus X^s(\epsilon)$ over $X^s(\epsilon)$, i.e.,

$$h : X^s(\epsilon) \to X^u(\epsilon) \oplus X^s(\epsilon)$$
$$(m, x^s) \to (m, x^u + x^s).$$

Again, with an abuse of notation, we write

$$x^u = h(m, x^s).$$

The graph of h is denoted by

$$\mathrm{gr}(h) = \{(m, x^u + x^s) \in X^u(\epsilon) \oplus X^s(\epsilon) : x^u = h(m, x^s)\}.$$

Define, for fixed $\mu \in (0, 1)$, $\epsilon > 0$, $\delta > 0$ and $\tilde{\epsilon} > 0$ the space of Lipschitz sections

$$\begin{aligned}
\Gamma^{cs} &= \Gamma^{cs}(\epsilon, \mu, \delta, \tilde{\epsilon}) \\
&\equiv \Big\{ h : \overline{X^s(\epsilon)} \to \overline{X^u(\epsilon)} \oplus \overline{X^s(\epsilon)} : \ \mathrm{gr}(h) \subset K_g'(\epsilon, 1/\mu, \delta) \\
&\qquad \text{and for all } (m, x^s) \in \overline{X^s(\epsilon)} \text{ if} \\
&\qquad \Theta^{-1}(m + x^s + h(m, x^s) + x_1^u + x_1^s + x_1^c) \in \ \mathrm{gr}(h) \\
&\qquad \text{for } x_1^\alpha \in \overline{X^\alpha(\tilde{\epsilon})}, \alpha = u, s, c, \text{ then} \\
&\qquad |x_1^u| \le \mu(|x_1^s| + |x_1^c|) \Big\}.
\end{aligned}$$

We may define a norm on Γ^{cs} by

$$||h|| = \sup\{|h(m, x^s)| \; : \; (m, x^s) \in \overline{X^s(\epsilon)}\},$$

which makes Γ^{cs} into a complete metric space. Note that from Lemma 4.5, there exist positive constants ϵ^* and $\tilde{\epsilon}^*$ such that if $\epsilon < \epsilon^*$ and $\tilde{\epsilon} < \tilde{\epsilon}^*$ then Γ^{cs} is not empty, in fact, the zero section is in Γ^{cs}.

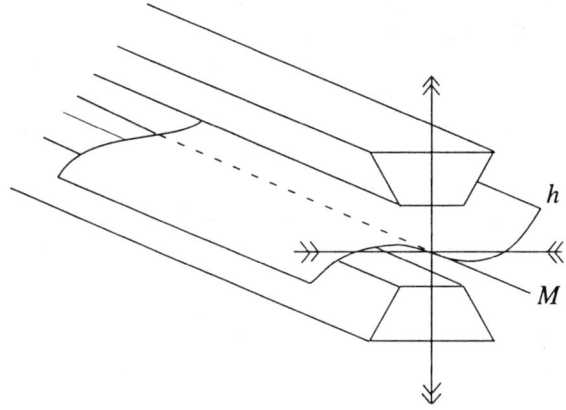

Figure 6. A Lipschitz Section, $h \in \Gamma^{cu}$

Before we state the main theorem of this section, we want to introduce two fundamental lemmas which give the Lipschitz properties of each section in Γ^{cs} under two different coordinate systems which were introduced in Section 4.

Lemma 7.1. *Let $\mu \in (0, 1)$ and $\rho \in (1, 2)$. For each $\tilde{\epsilon} < \tilde{\epsilon}^*$ there exists $\epsilon^* > 0$ such that for each $\epsilon \in (0, \epsilon^*)$ and for all $h \in \Gamma^{cs}$ and $m_0 \in M$, if*

$$g : (M \cap B(m_0, \rho\epsilon)) \times X^s_{m_0}(\rho^{-1}\epsilon) \to X^u_{m_0}$$

is defined by

$$\Pi^u_m g(m, \tilde{x}^s) = h(m, \Pi^s_m \tilde{x}^s) \tag{7.2}$$

then

$$|g(m_1, \tilde{x}^s_1) - g(m_2, \tilde{x}^s_2)| \leq \rho\mu(|m_1 - m_2| + |\tilde{x}^s_1 - \tilde{x}^s_2|) \tag{7.3}$$

Note that g is the local Lipschitz representative of h under the third coordinate system introduced in Section 4 by trivializing the bundle.

The next lemma gives the the local coordinate representative of h under the second coordinate system (Cartesian coordinate system) introduced in section 4.

Lemma 7.2. *Let* $\mu \in (0,1)$ *and* $\rho \in (1,2)$. *Then for each* $\tilde{\epsilon} < \tilde{\epsilon}^*$ *there exists* $\epsilon^* = \epsilon^*(\tilde{\epsilon})$ *and* $\delta^* = \delta^*(\epsilon) < \epsilon$ *such that for each* $\epsilon < \epsilon^*$ *and* $\delta < \delta^*$ *if* $h \in \Gamma^{cs}(\epsilon, \mu, \delta, \tilde{\epsilon})$ *and* $m_0 \in M$, *then there exists*

$$f : X^s_{m_0}(\rho^{-2}\epsilon) \times X^c_{m_0}(\epsilon) \to X^u_{m_0}(\epsilon)$$

such that

$$|f(x^s_1, x^c_1) - f(x^s_2, x^c_2)| \le \rho\mu(|x^s_1 - x^s_2| + |x^c_1 - x^c_2|) \tag{7.4}$$

and

$$\Theta^{-1}(m_0 + x^s_0 + x^c_0 + f(x^s_0, x^c_0)) \in gr(h)$$

for all $(x^s_0, x^c_0) \in X^s_{m_0}(\rho^{-2}\epsilon) \times X^c_{m_0}(\epsilon)$. *Furthermore, for all* $m \in M \cap B(m_0, \rho^{-1}\epsilon)$ *and* $x^s \in X^s_m(\rho^{-3}\epsilon)$,

$$m + x^s + h(m, x^s) = m_0 + x^s_0 + x^c_0 + f(x^s_0, x^c_0) \tag{7.5}$$

for some $(x^s_0, x^c_0) \in X^s_{m_0}(\rho^{-2}\epsilon) \times X^c_{m_0}(\epsilon)$.

The proofs of Lemma 7.1 and Lemma 7.2 follow the same lines as Lemma 6.1 and Lemma 6.2. We omit them.

Theorem 7.3. *Let* $\lambda_1 \in (\lambda, 1)$ *and* $\mu \in (0, 1)$. *Then there exist positive constants* $\tilde{\epsilon}^*$, $\epsilon^* = \epsilon^*(\tilde{\epsilon})$, $\delta^* = \delta^*(\epsilon) < \epsilon$ *and* $\sigma = \sigma(\epsilon, \delta)$ *such that if* $\tilde{\epsilon} < \tilde{\epsilon}^*$, $\epsilon < \epsilon^*$, $\delta < \delta^*$, *and* $\tilde{T}$ *satisfies* $\|\tilde{T} - T^t\|_1 < \sigma$, *then* $\tilde{T}$ *has a Lipschitz center-stable manifold*

$$\tilde{W}^{cs}(\epsilon) = \Theta(gr(\tilde{h}^{cs})),$$

where $\tilde{h}^{cs} \in \Gamma^{cs}(\epsilon, \mu, \delta, \tilde{\epsilon})$.

The proof of this theorem is built on the following proposition and lemmas. Throughout this section we shall assume that λ_1 and μ satisfy the requirements stated in Theorem 7.3 and the constant $\rho \in (1, 1/\sqrt{\lambda})$ will be that used in Lemmas 7.1 and 7.2. Note that the local coordinate representatives of $\tilde{h}^{cs}$, g and f, have Lipschitz constant $\rho\mu$

The next proposition is the core of this section, resulting in the graph transform.

Proposition 7.4. *There exist positive constants* $\tilde{\epsilon}^*$, $\epsilon^* = \epsilon^*(\tilde{\epsilon})$, $\delta^* = \delta^*(\epsilon) < \epsilon$ *and* $\sigma = \sigma(\epsilon, \delta)$ *such that if* $\tilde{\epsilon} < \tilde{\epsilon}^*$, $\epsilon < \epsilon^*$, $\delta < \delta^*$, *and* $\tilde{T}$ *satisfies* $\|\tilde{T} - T^t\|_1 < \sigma$, *then for each* $h \in \Gamma^{cs}$ *there exists a unique* $\tilde{h} \in \Gamma^{cs}$ *such that*

$$\tilde{T}(\Theta(gr(\tilde{h}))) \subset \Theta(gr(h))$$

This will be proved through a sequence of lemmas.

Lemma 7.5. *There exist positive constants $\tilde{\epsilon}^*$, $\epsilon^* = \epsilon^*(\tilde{\epsilon})$, $\delta^* = \delta^*(\epsilon) < \epsilon$ and $\sigma = \sigma(\epsilon, \delta)$ such that if $\tilde{\epsilon} < \tilde{\epsilon}^*$, $\epsilon < \epsilon^*$, $\delta < \delta^*$, and $\tilde{T}$ satisfies $\|\tilde{T} - T^t\|_1 < \sigma$, then for all $h \in \Gamma^{cs}(\epsilon, \mu, \delta, \tilde{\epsilon})$ and $(m_0, x_0^s) \in \overline{X^s(\epsilon)}$, there exists a unique $x_0^u \in \overline{X_{m_0}^u(\epsilon)}$ such that*

$$\tilde{T}(m_0 + x_0^u + x_0^s) \in \Theta(gr(h)). \tag{7.6}$$

Proof. Let $h \in \Gamma^{cs}$ and $(m_0, x_0^s) \in \overline{X^s(\epsilon)}$. Set $\bar{m} = T^t(m_0)$. For any $x^u \in \overline{X_{m_0}^u(\epsilon)}$, by Lemma 5.4, we write $\tilde{T}(m_0 + x^u + x_0^s)$ as

$$\tilde{T}(m_0 + x^u + x_0^s) = \bar{m} + \bar{x}^u + \bar{x}^s + \bar{x}^c = m_1 + x_1^u + x_1^s \tag{7.7}$$

where $x_1^\alpha \in X^\alpha(\epsilon_2)$ for $\alpha = u, s$. To show Lemma 7.5, it is enough to prove that there is a unique $x^u \in \overline{X_{m_0}^u(\epsilon)}$ such that $\bar{x}^u = f(\bar{x}^s, \bar{x}^c)$. To find such x^u, we consider a map ξ from $\overline{X_{m_0}^u(\epsilon)}$ to $X_{m_0}^u$ defined by

$$\xi(x^u) = x^u - \left(DT^t(m_0)\big|_{X_{m_0}^u} \right)^{-1} (\bar{x}^u - f(\bar{x}^s, \bar{x}^c))$$

We claim that ξ is a contraction on $\overline{X_{m_0}^u(\epsilon)}$.

We first prove that ξ is well-defined. Note that $DT^t(m_0) : X_{m_0}^u \to X_{\bar{m}}^u$ has a bounded inverse from hypothesis (H3). It is enough to show that there exists ϵ^*, $\delta^* = \delta^*(\epsilon) < \epsilon$ and σ such that if $\epsilon < \epsilon^*$, $\delta < \delta^*$, and $\tilde{T}$ satisfies $\|\tilde{T} - T^t\|_1 < \sigma$, then $(\bar{x}^s, \bar{x}^c) \in \overline{X_{\bar{m}}^s(\rho^{-2}\epsilon)} \times \overline{X^c(\epsilon)}$.

Applying the projection $\Pi_{\bar{m}}^s$ to (7.7) we have

$$\bar{x}^s = \Pi_{\bar{m}}^s(\tilde{T}(m_0 + x^u + x_0^s) - \bar{m}). \tag{7.8}$$

Using the Taylor expansion of $T^t(m_0 + x^u + x_0^s)$ at m_0 and hypothesis (H3), we obtain

$$\begin{aligned} |\bar{x}^s| &\leq \|\Pi_{\bar{m}}^s\| \, \|\tilde{T} - T^t\|_0 + |\Pi_{\bar{m}}^s DT^t(m_0)(x^u + x_0^s)| + O(\epsilon)(|x^u| + |x_0^s|) \\ &\leq C\sigma + \|DT^t(m_0)\big|_{X_{m_0}^s}\| \, |x_0^s| + O(\epsilon)\epsilon \\ &\leq (\lambda + O(\epsilon))\epsilon + C\sigma < \rho^{-2}\epsilon, \end{aligned} \tag{7.9}$$

provided $\rho < 1/\sqrt{\lambda}$ and ϵ^* and σ are sufficiently small.

Similarly, we have

$$\bar{x}^c = \Pi_{\bar{m}}^c(\tilde{T}(m_0 + x^u + x_0^s) - \bar{m})$$

and

$$\begin{aligned} |\bar{x}^c| &\leq C\|\tilde{T} - T^t\|_0 + O(\epsilon)\epsilon \\ &\leq C\sigma + O(\epsilon)\epsilon < \epsilon \end{aligned} \tag{7.10}$$

by choosing smaller ϵ^* and σ if necessary. Therefore, ξ is well-defined.

Next we show that $\xi : X_{m_0}^u(\epsilon) \to X_{m_0}^u$ is a contraction.

For $x_1^u, x_2^u \in X_{m_0}^u(\epsilon)$, we write

$$\tilde{T}(m_0 + x_i^u + x_0^s) = \bar{m} + \bar{x}_i^u + \bar{x}_i^s + \bar{x}_i^c, \ i = 1, 2. \tag{7.11}$$

For $\alpha = s, c$, estimate

$$
\begin{aligned}
|\bar{x}_1^\alpha - \bar{x}_2^\alpha| &= |\Pi_{\bar{m}}^\alpha(\tilde{T}(m_0 + x_1^u + x_0^s) - \tilde{T}(m_0 + x_2^u + x_0^s))| \\
&= |\Pi_{\bar{m}}^\alpha \left(\int_0^1 D\tilde{T}(m_0 + \tau x_1^u + (1-\tau)x_2^u + x_0^s)(x_1^u - x_2^u)d\tau \right)| \\
&\leq \|\Pi_{\bar{m}}^\alpha\| \|\tilde{T} - T^t\|_1 |x_1^u - x_2^u| \\
&\quad + |\Pi_{\bar{m}}^\alpha \int_0^1 DT^t(m_0 + \tau x_1^u + (1-\tau)x_2^u + x_0^s)(x_1^u - x_2^u)d\tau| \\
&\leq \left(C\sigma + O(\epsilon) \right)|x_1^u - x_2^u|.
\end{aligned}
\tag{7.12}
$$

In order to show ξ is a contraction we also need to estimate

$$
\begin{aligned}
&|x_1^u - x_2^u - (DT^t(m_0)|_{X_{m_0}^u})^{-1}(\bar{x}_1^u - \bar{x}_2^u)| \\
&= |(DT^t(m_0)|_{X_{m_0}^u})^{-1}\left(DT^t(m_0)(x_1^u - x_2^u) \right. \\
&\quad \left. - \Pi_{\bar{m}}^u \int_0^1 D\tilde{T}(m_0 + \tau x_1^u + (1-\tau)x_2^u + x_0^s)(x_1^u - x_2^u)d\tau \right)| \\
&\leq \left(C\sigma + O(\epsilon) \right)|x_1^u - x_2^u|.
\end{aligned}
\tag{7.13}
$$

Thus, from (7.4), (7.12) and (7.13) we obtain

$$
\begin{aligned}
&|\xi(x_1^u) - \xi(x_2^u)| \\
&= |x_1^u - x_2^u - \left(DT^t(m_0)|_{X_{m_0}^u} \right)^{-1}(\bar{x}_1^u - \bar{x}_2^u - f(\bar{x}_1^s, \bar{x}_1^c) + f(\bar{x}_2^s, \bar{x}_2^c))| \\
&\leq \left(C\sigma + O(\epsilon) \right)|x_1^u - x_2^u|.
\end{aligned}
\tag{7.14}
$$

This shows that ξ is a contraction by choosing ϵ^* and σ sufficiently small. To complete the proof, we have to show ξ maps $\overline{X_{m_0}^u(\epsilon)}$ into itself. We first estimate $\xi(0)$. Again using (7.7), we have

$$\bar{x}^u = \Pi_{\bar{m}}^u(\tilde{T}(m_0 + x_0^s) - T^t(m_0)),$$

and so

$$|\bar{x}^u| \leq C\sigma + O(\epsilon)\epsilon.$$

Hence, from (7.4), (7.9) and (7.10)

$$|\xi(0)| = |(DT^t(m_0)|_{X_{m_0}^u})^{-1}(\bar{x}^u - f(\bar{x}^s, \bar{x}^c)|$$
$$\leq \lambda(C\sigma + O(\epsilon)\epsilon + |f(0,0)| + \rho\mu(|\bar{x}^s| + |\bar{x}^c|))$$
$$\leq \lambda(C\sigma + O(\epsilon)\epsilon + \delta) < \lambda\epsilon$$

provided ϵ^*, δ^* and σ are sufficiently small.

For any $x^u \in X_{m_0}^u(\epsilon)$, from (7.14) it follows that

$$|\xi(x^u)| \leq |\xi(0)| + (C\sigma + O(\epsilon))\epsilon$$
$$< (\lambda\epsilon + C\sigma + C(\epsilon))\epsilon$$
$$< \epsilon$$

by choosing smaller ϵ^* and σ if necessary. This proves the claim that ξ is a contraction from $\bar{X}_{m_0}^u(\epsilon)$ into $X_{m_0}^u(\epsilon)$. By the contraction mapping theorem, ξ has a unique fixed point $x^u \in X_{m_0}^u(\epsilon)$, thus the corresponding $(\bar{x}^u, \bar{x}^s, \bar{x}^c)$ satisfies $\bar{x}^u = f(\bar{x}^u, \bar{x}^c)$. The proof of this lemma is complete.   □

We notice that the uniqueness of the fixed point of ξ in $\overline{X_{m_0}^s(\epsilon)}$ does not necessarily imply the uniqueness of the point x_0^u in (7.6). But we shall see this is a consequence of the following lemma.

Let $(m_0, x_0^u + x_0^s) \in \overline{X^u(\epsilon)} \oplus \overline{X^s(\epsilon)}$ such that

$$\tilde{T}(m_0 + x_0^u + x_0^s) \in \Theta(\mathrm{gr}(h)).$$

Consider a perturbation of $m_0 + x_0^u + x_0^s$

$$m_0 + x_0^u + x_0^s + x^u + x^s + x^c \in \overline{X^u(\epsilon)} \oplus \overline{X^s(\epsilon)}$$

where $x^\alpha \in \overline{X_{m_0}^u(\tilde{\epsilon})}$.

Lemma 7.6. *There exist positive constants $\tilde{\epsilon}^*$, $\epsilon^* = \epsilon^*(\tilde{\epsilon})$, $\delta^* = \delta^*(\epsilon) < \epsilon$ and $\sigma = \sigma(\epsilon, \delta)$ such that for $\tilde{\epsilon} < \tilde{\epsilon}^*$, $\epsilon < \epsilon^*$, $\delta < \delta^*$, and $\tilde{T}$ satisfying $||\tilde{T} - T^t||_1 < \sigma$, if*

$$\tilde{T}(m_0 + x_0^u + x_0^s + x^u + x^s + x^c) \in \Theta(gr(h))$$

then

$$|x^u| \leq \mu(|x^s| + |x^c|).$$

Proof. We prove this by contradiction. Suppose that

$$|x^u| > \mu(|x^s| + |x^c|).$$ (7.15)

Write

$$m_0 + x_0^u + x_0^s + x^u + x^s + x^c \tag{7.16}$$
$$= \bar{m}_0 + \bar{x}_0^u + \bar{x}_0^s,$$

where $\bar{x}_0^\alpha \in \overline{X_{\bar{m}_0}^\alpha(\epsilon)}$. Let

$$\tilde{T}(m_0 + x_0^u + x_0^s) = m_1 + x_1^u + x_1^s, \tag{7.17}$$

and

$$\tilde{T}(\bar{m}_0 + \bar{x}_0^u + \bar{x}_0^s) = \bar{m}_1 + \bar{x}_1^u + \bar{x}_1^s \tag{7.18}$$
$$= m_1 + x_1^u + x_1^s + \tilde{x}^u + \tilde{x}^s + \tilde{x}^c.$$

Let $\epsilon_1 < \frac{1}{2C}$, where C is the constant given in Lemma 4.5. Then applying Lemma 4.4, there exits β^* such that (4.12) holds with ϵ_1 instead of ϵ. Let η and r satisfy (4.14). By choosing ϵ^* and $\tilde{\epsilon}$ sufficiently small and using (7.16), $\bar{m}_0$ is in an η-neighborhood of m_0. Thus, applying Lemma 4.5 to the points m_0 and (7.16), we obtain

$$|m_0 - \bar{m}_0| \le \frac{1}{1 - C\epsilon_1} \, |x^c| \le 2\tilde{\epsilon},$$

and for $\alpha = u, s$,

$$|x_0^\alpha + x^\alpha - \left(\Pi_{\bar{m}_0}^\alpha\big|_{X_{m_0}^\alpha}\right)^{-1} \bar{x}_0^\alpha|$$
$$\le C\epsilon_1 |m_0 - \bar{m}_0|$$
$$\le C\epsilon_1 \tilde{\epsilon}.$$

This yields

$$|x^\alpha| \le |x_0^\alpha| + \frac{1}{1 - \eta} |\bar{x}_0^\alpha| + C\epsilon_1 \tilde{\epsilon}$$
$$\le C(\epsilon + \epsilon_1 \tilde{\epsilon})$$

where (4.5) and (4.14) are used to bound $\left\|\left(\Pi_{\bar{m}_0}^\alpha\big|_{X_{m_0}^\alpha}\right)^{-1}\right\|$. Thus, using the assumption (7.15), we find

$$|x^c| \le \frac{1}{\mu} C(\epsilon + \epsilon_1 \tilde{\epsilon}).$$

Set $\gamma(\tau) = m_0 + x_0^u + x_0^s + \tau(x^u + x^s + x^c)$. Then from (7.16)–(7.18), it follows that

$$|\tilde{x}^\alpha| = \left|\Pi_{m_1}^\alpha \left(\tilde{T}(m_0 + x_0^u + x_0^s + x^u + x^s + x^c)\right.\right.$$
$$\left.\left. - \tilde{T}(m_0 + x_0^u + x_0^s)\right)\right|$$
$$= \left|\Pi_{m_1}^\alpha \int_0^1 D\tilde{T}(\gamma(\tau))(x^u + x^s + x^c)d\tau\right|$$
$$\le \frac{1}{\mu} C(\epsilon + \epsilon_1 \tilde{\epsilon}) < \tilde{\epsilon}, \tag{7.19}$$

provided that $\epsilon_1 < \frac{\mu}{2C}$, $\epsilon < \frac{\mu\bar{\epsilon}}{2C}$. Since $m_1 + x_1^u + x_1^s \in \Theta(\mathrm{gr}(h))$ and

$$m_1 + x_1^u + x_1^s + \tilde{x}^u + \tilde{x}^s + \tilde{x}^c \in \Theta(\mathrm{gr}(h)),$$

$h \in \Gamma^{cs}$ implies

$$|\tilde{x}^\mu| \le \mu(|\tilde{x}^s| + |\tilde{x}^c|).$$

On the other hand, by Lemma 5.5, (7.15) implies

$$|\tilde{x}^u| > \mu(|\tilde{x}^s| + |\tilde{x}^c|),$$

and we have a contradiction. The proof is complete. $\qquad\square$

A direct consequence of this lemma is

Corollary 7.7. *There exist positive constants* $\bar{\epsilon}^*$, $\epsilon^* = \epsilon^*(\bar{\epsilon})$, $\delta^* = \delta^*(\epsilon) < \epsilon$ *and* $\sigma = \sigma(\epsilon, \delta)$ *such that if* $\bar{\epsilon} < \bar{\epsilon}^*$, $\epsilon < \epsilon^*$, $\delta < \delta^*$, *and* $\tilde{T}$ *satisfies* $\|\tilde{T} - T^t\|_1 < \sigma$, *then* x_0^u *in (7.6) is unique.*

Proof of Proposition 7.4. By Lemma 7.5 and Corollary 7.7, for $h \in \Gamma^{cs}$, we may define a section

$$\tilde{h}(m_0, x_0^s) = x_0^u$$

which induces a graph transform

$$\mathcal{F}^{cs}(h) = \tilde{h}.$$

Clearly, such a $\tilde{h}$ is unique. To complete the proof, we must show that $\tilde{h} \in \Gamma^{cs}$. By Lemma 7.6, it is enough to show that

$$(m_0, x_0^u + x_0^s) \in K_g'(\epsilon, 1/\mu, \delta).$$

We prove this by contradiction. Suppose that

$$(m_0, x_0^u + x_0^s) \in \mathrm{Interior}(K_g(\epsilon, 1/\mu, \delta)).$$

Note that $\tilde{T}(m_0 + x_0^u + x_0^s) \in X^u(\epsilon) \oplus X^s(\epsilon)$ from (7.6) and $h \in \Gamma^{cs}$. Thus, by Lemma 5.3, we have

$$\tilde{T}(m_0 + x_0^u + x_0^s) \in \mathrm{Interior}(K_g(\epsilon, 1/\mu, \delta)),$$

which contradicts the fact that

$$\tilde{T}(m_0 + x_0^u + x_0^s) = \bar{m} + \bar{x}^s + \bar{x}^c + f(\bar{x}^s, \bar{x}^c) \in \Theta(\mathrm{gr}(h)) \subset K_g'(\epsilon, 1/\mu, \delta).$$

This completes the proof. $\qquad\square$

We now have the graph transform $\mathcal{F}^{cs}$ defined on Γ^{cs}. We shall show that this is a contraction and that its fixed point gives the center-stable manifold.

Let $(m_0, x_0^u + x_0^s) \in X^u(\epsilon) \times X^s(\epsilon)$ be arbitrary such that

$$\tilde{T}(m_0 + x_0^u + x_0^s) \equiv \bar{m}_0 + \bar{x}_0^u + \bar{x}_0^s \in \Theta(X^u(\epsilon) \times X^s(\epsilon)). \tag{7.20}$$

Let $x_1^u \in X_{m_0}^u(\epsilon)$ be such that

$$\tilde{T}(m_0 + x_1^u + x_0^s) = \bar{m}_1 + \bar{x}_1^s + \bar{x}_1^u \in \Theta(\mathrm{gr}(h)), \tag{7.21}$$

for some $h \in \Gamma^{cs}$, i.e., $\bar{x}_1^u = h(\bar{m}_1, \bar{x}_1^s)$.

Lemma 7.8. *There exist positive constants* $\tilde{\epsilon}^*$, $\epsilon^* = \epsilon^*(\tilde{\epsilon})$, $\delta^* = \delta^*(\epsilon) < \epsilon$ *and* $\sigma = \sigma(\epsilon, \delta)$ *such that if* $\tilde{\epsilon} < \tilde{\epsilon}^*$, $\epsilon < \epsilon^*$, $\delta < \delta^*$, *and* $\tilde{T}$ *satisfies* $\|\tilde{T} - T^t\|_1 < \sigma$, *then*

$$|x_0^u - x_1^u| \leq \lambda_1 |\bar{x}_0^u - h(\bar{m}_0, \bar{x}_0^s)|$$

Proof. We first write (7.21) as

$$\tilde{T}(m_0 + x_1^u + x_0^s) = \bar{m}_1 + \bar{x}_1^u + \bar{x}_1^s$$
$$= \bar{m}_0 + \hat{x}_1^u + \bar{x}_0^s + x^u + x^s + x^c,$$

where $\bar{x}_1^\alpha \in X_{\bar{m}_1}^\alpha(\epsilon)$ for $\alpha = u, s$, $\bar{x}_0^s \in X_{\bar{m}_0}^s(\epsilon)$, $\hat{x}_1^u = h(\bar{m}_0, \bar{x}_0^s)$ and $x^\alpha \in X_{\bar{m}_0}^\alpha$. We want to estimate $|x_1^u - x_0^u|$ in terms of $|\hat{x}_1^u - \bar{x}_0^u|$.

From (7.20) and (7.21), by elementary computations, we find

$$|\bar{m}_i - T^t(m_0)| \leq \sigma + O(\epsilon), \text{ for } i = 0, 1,$$

and $|x^\alpha| \leq C\sigma + O(\epsilon) < \tilde{\epsilon}$ by choosing ϵ^* and σ small enough.

Thus, for $\alpha = s, c$,

$$|x^\alpha| = \left|\Pi_{\bar{m}_0}^\alpha(\tilde{T}(m_0 + x_1^u + x_0^s) - \tilde{T}(m_0 + x_0^u + x_0^s))\right| \tag{7.22}$$
$$\leq (C\sigma + O(\epsilon))|x_1^u - x_0^u|$$

and

$$|x^u| \leq \mu(|x^s| + |x^c|)$$
$$\leq (C\sigma + O(\epsilon))|x_1^u - x_0^u|, \tag{7.23}$$

since $h \in \Gamma^{cs}$.

Similarly, from (7.20) and (7.21), using (H3) we obtain

$$|\hat{x}_1^u + x^u - \bar{x}_0^u|$$
$$= \left|\Pi_{\bar{m}_0}^u(\tilde{T}(m_0 + x_1^u + x_0^s) - \tilde{T}(m_0 + x_0^u + x_0^s)\right|$$
$$\geq \left|\Pi_{T^t(m_0)}^u DT^t(m_0)(x_1^u - x_0^u)\right| - (C\sigma + O(\epsilon))|x_1^u - x_0^u|$$
$$\geq (\lambda^{-1} - C\sigma - O(\epsilon))|x_1^u - x_0^u|.$$

Together with (7.23), the above implies

$$|\hat{x}_1^u - \bar{x}_0^u|$$
$$\geq (\lambda^{-1} - C\sigma - O(\epsilon))|x_1^u - x_0^u|$$
$$\geq \lambda_1^{-1}|x_1^u - x_0^u|,$$

provided ϵ^* and σ are sufficiently small. This completes the proof. $\quad\square$

Proposition 7.9. *There exist positive constants $\tilde{\epsilon}^*$, $\epsilon^* = \epsilon^*(\tilde{\epsilon})$, $\delta^* = \delta^*(\epsilon) < \epsilon$ and $\sigma = \sigma(\epsilon, \delta)$ such that if $\tilde{\epsilon} < \tilde{\epsilon}^*$, $\epsilon < \epsilon^*$, $\delta < \delta^*$, and $\tilde{T}$ satisfies $\|\tilde{T} - T^t\|_1 < \sigma$, then $\mathcal{F}^{cs} : \Gamma^{cs}(\epsilon, \mu, \delta, \tilde{\epsilon}) \to \Gamma^{cs}(\epsilon, \mu, \delta, \tilde{\epsilon})$ is a contraction.*

Proof. For h_0, $h_1 \in \Gamma^{cs}$ and $(m_0, x_0^s) \in X^s(\epsilon)$, let $x_i^u = \mathcal{F}^{cs}(h_i)(m_0, x_0^s)$ for $i = 0, 1$. From the definition of $\mathcal{F}^{cs}$, there exist $(\bar{m}_i, \bar{x}_i^s) \in X^s(\epsilon), i = 0, 1$, such that

$$\tilde{T}(m_0 + x_0^u + x_0^s) = \bar{m}_0 + \bar{x}_0^s + h_0(\bar{m}_0, \bar{x}_0^s)$$

and

$$\tilde{T}(m_0 + x_1^u + x_0^s) = \bar{m}_1 + \bar{x}_1^s + h_1(\bar{m}_1, \bar{x}_1^s).$$

Then applying Lemma 7.8, we obtain

$$|x_0^u - x_1^u| \le \lambda_1 \|h_1 - h_0\|.$$

This completes the proof.   □

Proof of Theorem 7.3. By Lemma 7.6, $\mathcal{F}^{cs}$ is a contraction, hence, it has a unique fixed point $\tilde{h}^{cs} \in \Gamma^{cs}$,

$$\mathcal{F}^{cs}(\tilde{h}^{cs}) = \tilde{h}^{cs}.$$

The definition of $\mathcal{F}^{cs}$ gives

$$\tilde{T}(\Theta(\text{gr }(\tilde{h}^{cs}))) \subset \Theta(\text{gr }(\tilde{h}^{cs}))$$

which yields the positive invariance of $\Theta(\text{gr}(\tilde{h}^{cs}))$. This completes the proof.   □

Similar to Proposition 6.10, we define by induction a sequence of sets $\mathcal{A}_{-k}, k = 1, 2, \cdots$ as follows

$$\mathcal{A}_{-k} = \tilde{T}^{-1}(\mathcal{A}_{1-k}) \cap \Theta(X^u(\epsilon) \oplus X^s(\epsilon)), \text{ for } k \ge 1,$$

where $\mathcal{A}_0 = \Theta(X^u(\epsilon) \oplus X^s(\epsilon))$. In the same fashion as in Proposition 6.9, one may prove

Proposition 7.10.
$$\tilde{W}^{cs}(\epsilon) = \cap_{k=1}^{\infty} \mathcal{A}_{-k}$$

Using Theorem 7.3, there exists a center-stable manifold for the time-t map T^t given by

$$W^{cs}(\epsilon) = \Theta(gr(h^{cs}))$$

where $h^{cs} \in \Gamma^{cs}$. The next result states that the perturbed center-stable manifold $\tilde{W}^{cs}(\epsilon)$ is close to the unperturbed one, $W^{cs}(\epsilon)$.

Proposition 7.11. *There exists ϵ^* such that for $\epsilon < \epsilon^*$,*

$$\|\tilde{h}^{cs} - h^{cs}\| \to 0, \quad as \quad \|\tilde{T} - T^t\|_0 \to 0.$$

Proof. We denote the ϵ^* in Theorem 7.3 temporarily by ϵ_1^*. Let $\epsilon_2 < \epsilon_1^*$ be fixed. By Lemma 4.3, we may choose ϵ^* sufficiently small such that for $\epsilon < \epsilon^*$, $(m_1, x_1^u + x_1^s) \in X^u(\epsilon) \oplus X^s(\epsilon)$, and $x \in X$ if $|m_1 + x_1^u + x_1^s - x| < \epsilon$, then $x = m_2 + x_2^u + x_2^s$, where $x_2^\alpha \in X_{m_2}^\alpha(\epsilon_2)$.

Observe that Proposition 7.10 implies that $\tilde{W}^{cs}(\epsilon) \subset \tilde{W}^{cs}(\epsilon_2)$ for all $\epsilon < \epsilon_2 < \epsilon_1^*$. In other words, if we denote $\tilde{h}^{cs}$ by $\tilde{h}_\epsilon^{cs}$, then we have $\tilde{h}_\epsilon^{cs} = \tilde{h}_{\epsilon_2}^{cs}$ on $X^u(\epsilon)$.

Let $\mathcal{F}^{cs}$ denote the graph transform for the time-t map T^t and $\tilde{\mathcal{F}}^{cs}$ that for $\tilde{T}$. From Proposition 7.9, $\mathcal{F}^{cs}$ is a contraction in $\Gamma^{cs}(\epsilon_2, \mu, \delta, \tilde{\epsilon})$ and has the fixed point h^{cs}. Thus, given small $\mathcal{E} > 0$, there is a positive integer k such that

$$\left\|\left(\mathcal{F}^{cs}\right)^k(\tilde{h}^{cs}) - h^{cs}\right\| \leq \mathcal{E}. \tag{7.24}$$

We may also choose k such that $\lambda_1^k \leq \mathcal{E}$. For such fixed k, it is easy to see that there exists $\beta > 0$ such that if $\|\tilde{T} - T^t\|_0 \leq \beta$ then for $i = 1, 2, \cdots, k$

$$\|\tilde{T}^i - (T^t)^i\|_0 \leq \mathcal{E}. \tag{7.25}$$

Let ϵ^* also satisfy $\epsilon^* < \epsilon_2$. For each $(m_0, x_0^u) \in X^s(\epsilon)$, let

$$\hat{x}_0^u = \left(\mathcal{F}^{cs}\right)^k \tilde{h}^{cs}(m_0, x_0^s)$$

and

$$x_0^u = \left(\tilde{\mathcal{F}}^{cs}\right)^k \tilde{h}^{cs}(m_0, x_0^s).$$

For $0 \leq i \leq k$, we write

$$(T^t)^i(m_0 + x_0^u + x_0^s) = m_i + x_i^u + x_i^s$$

and let

$$\hat{x}_i^u = \left(\mathcal{F}^{cs}\right)^{k-i} \tilde{h}^{cs}(m_i, x_i^s)$$

Using Lemma 7.8, we obtain

$$|\hat{x}_i^u - x_i^u| \leq \lambda_1 |\hat{x}_{i+1}^u - x_{i+1}^u|,$$

which yields

$$|\hat{x}_0^u - x_0^u| \leq \lambda_1^k |\hat{x}_k^u - x_k^u| \leq 2\epsilon_1 \mathcal{E}.$$

Therefore,

$$|\tilde{h}^{cs}(m_0, x_0^s) - h^{cs}(m_0, x_0^s)|$$

$$\leq |\left(\tilde{\mathcal{F}}^{cs}\right)^k(\tilde{h}^{cs})(m_0, x_0^s) - \left(\mathcal{F}^{cs}\right)^k(\tilde{h}^{cs})(m_0, x_0^s)| + |\left(\mathcal{F}^{cs}\right)^k(\tilde{h}^{cs})(m_0, x_0^s) - h^{cs}(m_0, x_0^s)|$$

$$\leq C\mathcal{E},$$

which completes the proof. $\qquad\qquad\square$

8. Smoothness of Center-Stable Manifolds.

We shall see that the center-stable manifold obtained in Section 7 indeed is C^1 by choosing ϵ^* and σ smaller than those in Theorem 7.1. More precisely, we have

Theorem 8.1. *Let* $\lambda_1 \in (\lambda, 1)$, $\rho \in (1, 1/\sqrt{\lambda})$, *and* $\mu \in (0, 1)$ *such that* $\mu\rho < 1/2$. *Then there exist positive constants* $\tilde{\epsilon}^*$, $\epsilon^* = \epsilon^*(\tilde{\epsilon})$, $\delta^* = \delta^*(\epsilon) < \epsilon$ *and* $\sigma = \sigma(\epsilon, \delta)$ *such that if* $\tilde{\epsilon} < \tilde{\epsilon}^*$, $\epsilon < \epsilon^*$, $\delta < \delta^*$, *and* $\tilde{T}$ *satisfies* $\|\tilde{T} - T^t\|_1 < \sigma$, *then* $\tilde{W}^{cs}(\epsilon)$ *is a* C^1 *manifold.*

The basic idea to show its smoothness is to find a candidate for the tangent bundle of this manifold, which is invariant under the linearization $D\tilde{T}$, then to prove it indeed is tangent to the manifold. The arguments are based on the use of Lipschitz jets, which is borrowed from [HPS]. Since the trivialization of the normal bundle of M is not available in a Banach space, the proof is more complicated than for finite dimensional systems. We first define a space of sections of the Lipschitz jet bundle, which is different from the jet spaces introduced in [HPS]. Then we construct a graph transform based on the linearization $D\tilde{T}$ and show that it has a unique fixed point which gives the tangent bundle of the center-stable manifold. A major difficulty in finding the fixed point is that the space of sections of the Lipschitz jet bundle is not complete. Finally, we prove that the tangent bundle is C^0.

Let Y and Z be Banach spaces. For $y_0 \in Y$ and $z_0 \in Z$, consider two local maps

$$g_i : U_i \to Z, \ g_i(y_0) = z_0, \ i = 1, 2$$

where U_i is a neighborhood of y_0.

Define

$$d(g_1, g_2) = \overline{\lim_{y \to y_0}} \frac{|g_1(y) - g_2(y)|}{|y - y_0|}.$$

We shall see that $d(g_1, g_2)$ defines a metric.

If $d(g_1, g_2) = 0$, we say that g_1 is equivalent to g_2. The equivalence class of all local maps equivalent to g_1 is called the Lipschitz jet of g_1 at y_0, which is denoted by $j_1 = [g_1]$. We use $J(Y, Z; y_0, z_0)$ to denote the set of all jets at y_0 carrying y_0 to z_0. For $j_1, j_2 \in J(Y, Z; y_0, z_0)$, we define

$$d(j_1, j_2) = \overline{\lim_{y \to y_0}} \frac{|g_1(y) - g_2(y)|}{|y - y_0|}$$

where g_1 and g_2 are representatives of j_1 and j_2 respectively. It is not hard to see that $d(j_1, j_2)$ does not depend on the choices of the representatives.

Consider the jet spaces

$J^b = \{j \in J(Y, Z; y_0, z_0) : \ d(j, [z_0]) < \infty\}$

$J^c = \{j \in J^b : \ j \text{ has a representative which is continuous in a neighborhood of } y_0\}$

$J^d = \{j \in J^b : \ j \text{ has a differentiable representative}\}$

$J^a = \{j \in J^b : \ j \text{ has an affine representative}\}$

Theorem on Lipschitz Jets. *Let $y_0 = 0$, $z_0 = 0$. Then J^b is a Banach space with norm $\|j\| = d(j, 0)$. The sets J^c, J^d and J^a are closed subspaces of J^b and*

$$J^b \supset J^c \supset J^d = J^a.$$

The above results are borrowed from [HPS].

For each $m \in M$. Set

$$J^b(m) = J^b(X_m^s \times X_m^c, X_m^u; 0, 0),$$
$$J^c(m) = J^c(X_m^s \times X_m^c, X_m^u; 0, 0),$$
$$J^d(m) = J^d(X_m^s \times X_m^c, X_m^u; 0, 0),$$
$$J^a(m) = J^a(X_m^s \times X_m^c, X_m^u; 0, 0).$$

Let $\theta = \mu\rho < 1/2$. For each fixed $m \in M$, we define a Lipschitz jet space

$$J_\theta^l(m) = J_\theta^l(X_m^s \times X_m^c, X_m^u; 0, 0)$$
$$= \{j \in J^b(m) : j \text{ has a representative } g : U \to X_m^u \text{ satisfying Lip } (g|_U) \leq \theta\}$$

where U is a neighborhood of 0 in $X_m^s \times X_m^c$ and

$$\text{Lip}(g|_U) = \sup\{\frac{|g(x^s, x^c) - g(\bar{x}^s, \bar{x}^c)|}{|x^s - \bar{x}^s| + |x^c - \bar{x}^c|} : (x^s, x^c), (\bar{x}^s, \bar{x}^c) \in U, (x^s, x^c) \neq (\bar{x}^s, \bar{x}^c)\}.$$

Let $J_\theta^l = J_\theta^l(X^s \times X^c, X^u; 0, 0)$ denote the Lipschitz jet bundle over M with fiber $J_\theta^l(m)$. Set $J_\theta^a = J_\theta^l \cap J^a$.

Consider all maps from $\tilde{W}^{cs}(\epsilon)$ to J_θ^l which map points $m + x^s + \tilde{h}^{cs}(m, x^s)$ to jets $j = j(m, x^s)$ in $J_\theta^l(X_m^s \times X_m^c, X_m^u; 0, 0)$.

Define

$$\Sigma_\theta^{cs,l} = \{\gamma : \tilde{W}^{cs}(\epsilon) \to J_\theta^l : \|\gamma\| < \infty\}$$

where

$$\|\gamma\| = \sup\{\|j(m, x^s)\| : m + x^s + \tilde{h}^{cs}(m, x^s) \in \tilde{W}^{cs}(\epsilon)\},$$

and we identify $j(m, x^s)$ with $\gamma(m + x^s + \tilde{h}^{cs}(m, x^s))$. Similarly, we may define $\Sigma_\theta^{cs,a}$.

For $m_1 + x_1^u + x_1^s \in \tilde{W}^{cs}(\epsilon)$, from the invariance of $\tilde{W}^{cs}(\epsilon)$, we may write

$$\tilde{T}(m_1 + x_1^u + x_1^s) = m_2 + x_2^u + x_2^s \tag{8.1}$$

where $x_2^\alpha \in X_{m_2}^\alpha(\epsilon)$ for $\alpha = u, s$ and $x_i^u = \tilde{h}^{cs}(m_i, x_i^s), i = 1, 2$.

Let $\gamma \in \Sigma_\theta^{cs,l}$ be fixed and set for $i = 1, 2$

$$j_i = \gamma(x_i), \tag{8.2}$$

where $x_i = m_i + x_i^u + x_i^s$. Let $g_2 : B_2^s(0, r_2) \times B_2^c(0, r_2) \to X_{m_2}^u$ be a Lipschitz representative of j_2 such that for $x^\alpha, \bar{x}^\alpha \in B_2^\alpha(0, r_2), \alpha = c, s$

$$|g_2(x^s, x^c) - g_2(\bar{x}^s, \bar{x}^c)| \leq \theta(|x^s - \bar{x}^s| + |x^c - \bar{x}^c|), \qquad (8.3)$$

where $B_2^\alpha(0, r_2)$ is the ball in $X_{m_2}^\alpha$ with radius $r_2 > 0$ centered at 0.

We shall construct $\tilde{\gamma}$ such that γ is the image of $\tilde{\gamma}$ under $D\tilde{T}$ in a certain sense, which will be stated precisely later. For the given g_2, we want to find a Lipschitz map $\tilde{g}_1$ defined on $B_1^s(0, r_1) \times B_1^c(0, r_1)$ for some $r_1 > 0$ such that for each $(x^s, x^c) \in B_1^s(0, r_1) \times B_1^c(0, r_1)$ there exists $(\bar{x}^s, \bar{x}^c) \in B_2^s(0, r_2) \times B_2^c(0, r_2)$ satisfying

$$D\tilde{T}(x_1)(x^s + x^c + \tilde{g}_1(x^s, x^c)) = \bar{x}^s + \bar{x}^c + g_2(\bar{x}^s, \bar{x}^c). \qquad (8.4)$$

To see this, for $(x^s, x^c) \in B_1^s(0, r_1) \times B_1^c(0, r_1)$, we define a map E from $B_1^u(0, r_1)$ to $X_{m_1}^u$ as follows

$$E(x^u)$$
$$= \left(\Pi_{m_2}^u D\tilde{T}(x_1)|_{X_{m_1}^u} \right)^{-1} (g_2(\Pi_{m_2}^s D\tilde{T}(x_1)(x^u + x^s + x^c), \Pi_{m_2}^c D\tilde{T}(x_1)(x^u + x^s + x^c))$$
$$- \Pi_{m_2}^u D\tilde{T}(x_1)(x^s + x^c)). \qquad (8.5)$$

We shall see that E is a contraction and its fixed point gives a solution of (8.4)

Before we study the properties of E, we state a technical lemma providing fundamental properties for $\tilde{T}$, which are inherited from the time-t map T^t.

Lemma 8.2. *There exist $\epsilon^* > 0$ and $\sigma > 0$ such that for $\epsilon < \epsilon^*$ if $x_1 \in \tilde{W}^{cs}(\epsilon)$ and $\|\tilde{T} - T^t\|_1 < \sigma$, then the following estimates hold*

(i) $\left\| \left(\Pi_{m_2}^u D\tilde{T}(x_1)|_{X_{m_1}^u} \right)^{-1} \right\| < \lambda$

(ii) $\left\| \left(\Pi_{m_2}^u D\tilde{T}(x_1)|_{X_{m_1}^u} \right)^{-1} \right\| \|\Pi_{m_2}^c D\tilde{T}(x_1)|_{X_{m_1}^c}\| < \lambda$

(iii) $\left\| \left(\Pi_{m_2}^c D\tilde{T}(x_1)|_{X_{m_1}^c} \right)^{-1} \right\| \|\Pi_{m_2}^s D\tilde{T}(x_1)|_{X_{m_1}^s}\| < \lambda.$

Proof. Let us first consider (i). Note that the hypothesis (H3) gives

$$\inf \left\{ |DT^t(m_1)x^u| \, : \, x^u \in X_{m_1}^u, |x^u| = 1, m_1 \in M \right\} > \frac{1}{\lambda} \qquad (8.6)$$

Let $\bar{m} = T^t(m_1)$, then

$$\left\| \left(\Pi_{\bar{m}}^u DT^t(m_1)|_{X_{m_1}^u} \right)^{-1} \right\| < \lambda.$$

From (8.1) and (H1), we obtain

$$|\bar{m} - m_2|$$
$$= |T^t(m_1) - \tilde{T}(m_1 + x_1^u + x_1^s) + x_2^u + x_2^s|$$
$$\leq |T^t(m_1) - T^t(m_1 + x_1^u + x_1^s)| + |T^t(m_1 + x_1^u + x_1^s) - \tilde{T}(m_1 + x_1^u + x_1^s)| + |x_2^u + x_2^s|$$
$$\leq C\epsilon + \sigma. \qquad (8.7)$$

We write

$$\Pi_{m_2}^u D\tilde{T}(m_1 + x_1^u + x_1^s)|_{X_{m_1}^u}$$
$$= \left(\Pi_{m_2}^u - \Pi_{\bar{m}}^u\right) D\tilde{T}(m_1 + x_1^u + x_1^s)|_{X_{m_1}^u}$$
$$+ \Pi_{\bar{m}}^u \left(D\tilde{T}(m_1 + x_1^u + x_1^s) - DT^t(m_1 + x_1^u + x_1^s)\right)|_{X_{m_1}^u}$$
$$+ \Pi_{\bar{m}}^u \left(DT^t(m_1 + x_1^u + x_1^s) - DT^t(m_1)\right)|_{X_{m_1}^u} + \Pi_{\bar{m}}^u DT^t(m_1)|_{X_{m_1}^u}.$$

Thus, from (H1), (H4), (8.6) and (8.7) it follows that

$$\inf\{|\Pi_{m_2}^u D\tilde{T}(m_1 + x_1^u + x_1^s)x^u| : x^u \in X_{m_1}^u, |x^u| = 1\}$$
$$\geq \inf\left\{|DT^t(m_1)x^u| : x^u \in X_{m_1}^u, |x^u| = 1\right\} - \|\Pi_{m_2}^u - \Pi_{\bar{m}}^u\|\|D\tilde{T}(m_1 + x_1^u + x_1^s)|_{X_{m_1}^u}\|$$
$$- \|\Pi_{\bar{m}}^u\|\|D\tilde{T} - DT^t\|_0 - \|\Pi_{\bar{m}}^u\|\|DT^t(m_1 + x_1^u + x_1^s) - DT^t(m_1)\|$$
$$\geq \inf\left\{|DT^t(m_1)x^u| : x^u \in X_{m_1}^u, |x^u| = 1\right\} - (C\sigma + O(\epsilon)) > \frac{1}{\lambda}$$

provided that ϵ^* and σ are sufficiently small. To complete the proof of (i) we must show that $\Pi_{m_2}^u D\tilde{T}(x_1)|_{X_{m_1}^u}$ is an isomorphism from $X_{m_1}^u$ onto $X_{m_2}^u$. We estimate the difference

$$\|\Pi_{m_2}^u D\tilde{T}(x_1)|_{X_{m_1}^u} - \Pi_{m_2}^u DT^t(m_1)|_{X_{m_1}^u}\|$$
$$\leq C\sigma + O(\epsilon).$$

Choose ϵ^* and σ sufficiently small such that m_2 is in an η-neighborhood of $\bar{m}$, where $\eta < \sqrt{2} - 1$. Thus from (4.5), we have $\Pi_{m_2}^u|_{X_{\bar{m}}^u}$ is an isomorphism. Furthermore, from that the fact that $DT^t(m_1)|_{X_{m_1}^u}$ is an isomorphism from $X_{m_1}^u$ onto $X_{\bar{m}}^u$ it follows that $\Pi_{m_2}^u DT^t(m_1)|_{X_{m_1}^u}$ is an isomorphism. Hence $\Pi_{m_2}^u D\tilde{T}(x_1)|_{X_{m_1}^u}$ is also an isomorphism as long as ϵ^* and σ are sufficiently small so (i) holds.

To see that (ii) and (iii) hold, we notice that the hypothesis (H3) also gives

$$\left\| \left(\Pi_{\bar{m}}^u DT^t(m_1)|_{X_{m_1}^u}\right)^{-1} \right\| \|\Pi_{\bar{m}}^c DT^t(m_1)|_{X_{m_1}^c}\| < \lambda$$

and

$$\left\| \left(\Pi_{\tilde{m}}^c DT^t(m_1)|_{X_{m_1}^c} \right)^{-1} \right\| \|\Pi_{\tilde{m}}^s DT^t(m_1)|_{X_{m_1}^s}\| < \lambda.$$

Thus one may prove (ii) and (iii) in the same fashion as (i).

This completes the proof   □

There is a simple estimate which we shall use frequently

$$\|\Pi_{m_2}^\alpha D\tilde{T}(m_1 + x_1^u + x_1^s) - \Pi_{\tilde{m}}^\alpha DT^t(m_1)\| \le C\sigma + O(\epsilon). \qquad (8.8)$$

In fact (8.8) follows from (H1), (H4) and (8.7), since

$$\|\Pi_{m_2}^\alpha D\tilde{T}(m_1 + x_1^u + x_1^s) - \Pi_{\tilde{m}}^\alpha DT^t(m_1)\|$$
$$\le \| \left(\Pi_{m_2}^\alpha - \Pi_{\tilde{m}}^\alpha \right) D\tilde{T}(m_1 + x_1^u + x_1^s)\|$$
$$+ \|\Pi_{\tilde{m}}^\alpha\|\|D\tilde{T}(m_1 + x_1^u + x_1^s) - DT^t(m_1 + x_1^u + x_1^s)\|$$
$$+ \|\Pi_{\tilde{m}}^\alpha\|\|DT^t(m_1 + x_1^u + x_1^s) - DT^t(m_1)\| \le C\sigma + O(\epsilon).$$

The next lemma shows that the map E given by (8.5) is well-defined and is a contraction.

Lemma 8.3. *There exist $\epsilon^* > 0$ and $\sigma > 0$ such that for $\epsilon < \epsilon^*$ and $\tilde{T}$ satisfying $\|\tilde{T} - T^t\|_1 < \sigma$ if $m_1 + x_1^u + x_1^s \in \tilde{W}^{cs}(\epsilon)$ and g_2 satisfies (8.3), then there exists $r_1^* = r_1^*(r_2) > 0$ such that for $r_1 < r_1^*$, $E : B_1^u(0, r_1) \to B_1^u(0, r_1)$ is a contraction.*

Proof. We first show that E is well-defined from $B_1^u(0, r_1)$ into itself. Observe that for $\alpha = s, c$, and $(x^u, x^s, x^c) \in B_1^u(0, r_1) \times B_1^s(0, r_1) \times B_1^c(0, r_1)$

$$|\Pi_{m_2}^\alpha D\tilde{T}(m_1 + x_1^u + x_1^s)(x^u + x^s + x^c)| \le C r_1 \le r_2$$

provided that $r_1^* < r_2/C$. Thus, $g_2(\Pi_{m_2}^s D\tilde{T}(x_1)(x^u + x^s + x^c), \Pi_{m_2}^c D\tilde{T}(x_1)(x^u + x^s + c^c))$ is well-defined. Note again that $m_1 + x_1^u + x_1^s$ is denoted by x_1. From Lemma 8.2 $\left(\Pi_{m_2}^u D\tilde{T}(x_1)|_{X_{m_1}^u} \right)^{-1}$ exists. Hence, $E(x^u)$ is defined for $x^u \in B_1^u(0, r_1)$.

Next, we show $|E(x^u)| \le r_1$. Using (8.3), (8.8) and Lemma 8.2, we obtain

$$|E(x_u)|$$
$$= | \left(\Pi_{m_2}^u D\tilde{T}(x_1)|_{X_{m_1}^u} \right)^{-1} \left(g_2(\Pi_{m_2}^s D\tilde{T}(x_1)(x^u + x^s + x^c), \Pi_{m_2}^c D\tilde{T}(x_1)(x^u + x^s + x^c)) \right.$$
$$\left. - \Pi_{m_2}^u D\tilde{T}(x_1)(x^s + x^c) \right)|$$
$$\le \| \left(\Pi_{m_2}^u D\tilde{T}(x_1)|_{X_{m_1}^u} \right)^{-1} \| \left(\theta \left(|\Pi_{m_2}^s D\tilde{T}(x_1)(x^u + x^s + x^c)| + |\Pi_{m_2}^c D\tilde{T}(x_1)(x^u + x^s + x^c)| \right) \right.$$
$$\left. + |\Pi_{m_2}^u D\tilde{T}(x_1)(x^s + x^c)| \right)$$
$$\le \| \left(\Pi_{m_2}^u D\tilde{T}(x_1)|_{X_{m_1}^u} \right)^{-1} \| \left(C\sigma + O(\epsilon) + \theta(\|\Pi_{m_2}^s D\tilde{T}(x_1)|_{X_{m_1}^s}\| + \|\Pi_{m_2}^c D\tilde{T}(x_1)|_{X_{m_1}^c}\|) \right) r_1$$
$$\le \left(C\sigma + O(\epsilon) + \theta\lambda(\lambda + 1) \right) r_1.$$

Since $0 < \lambda < 1$ and $\theta < 1/2$, by choosing ϵ^* and σ sufficiently small, we have that $C\sigma + O(\epsilon) + \theta\lambda(\lambda + 1) < 1$.

Finally, we show that E is a contraction. For $x^u, \bar{x}^u \in B_1^u(0, r_1)$, from (8.3), (8.5) and (8.8) and Lemma 8.2

$$
|E(x^u) - E(\bar{x}^u)|
$$

$$
= \left| \left(\Pi_{m_2}^u D\tilde{T}(x_1)|_{X_{m_1}^u} \right)^{-1} \left(g_2(\Pi_{m_2}^s D\tilde{T}(x_1)(x^u + x^s + x^c), \Pi_{m_2}^c D\tilde{T}(x_1)(x^u + x^s + x^c)) \right. \right.
$$

$$
\left. \left. - g_2(\Pi_{m_2}^s D\tilde{T}(x_1)(\bar{x}^u + x^s + x^c), \Pi_{m_2}^c D\tilde{T}(x_1)(\bar{x}^u + x^s + x^c)) \right) \right|
$$

$$
\leq \left\| \left(\Pi_{m_2}^u D\tilde{T}(x_1)|_{X_{m_1}^u} \right)^{-1} \right\| \theta \left(|\Pi_{m_2}^s D\tilde{T}(x_1)(x^u - \bar{x}^u)| + |\Pi_{m_2}^c D\tilde{T}(x_1)(x^u - \bar{x}^u)| \right)
$$

$$
\leq \lambda\theta \left(\|\Pi_{m_2}^s D\tilde{T}(x_1) - \Pi_{\bar{m}}^s DT^t(m_1)\| + \|\Pi_{m_2}^c D\tilde{T}(x_1) - \Pi_{\bar{m}}^c DT^t(m_1)\| \right) |x^u - \bar{x}^u|
$$

$$
\leq (C\sigma + O(\epsilon))|x^u - \bar{x}^u| \tag{8.9}
$$

which yields that E is a contraction by choosing ϵ^* and σ sufficiently small. This completes the proof. □

By the contraction mapping theorem, we obtain that for each $(x^s, x^c) \in B_1^s(0, r_1) \times B_1^c(0, r_1)$, E has a unique fixed point $x^u \in B_1^u(0, r_1)$, which defines a map from $B_1^s(0, r_1) \times B_1^c(0, r_1)$ to $B_1^u(0, r_1)$. We denote it by $x^u = \tilde{g}_1(x^s, x^c)$. Clearly $\tilde{g}_1$ satisfies (8.4). Furthermore, this function is Lipschitz with Lipschitz constant θ.

Lemma 8.4. *There exist $\epsilon^* > 0$ and $\sigma > 0$ such that for $\epsilon < \epsilon^*$ if $m_1 + x_1^u + x_1^s \in \tilde{W}^{cs}(\epsilon)$ and $\|\tilde{T} - T^t\|_1 < \sigma$, then for $(x^s, x^c), (\bar{x}^s, \bar{x}^c) \in B_1^s(0, r_1) \times B_1^c(0, r_1)$,*

$$
|\tilde{g}_1(x^s, x^c) - \tilde{g}_1(\bar{x}^s, \bar{x}^c)| \leq \theta \left(|x^s - \bar{x}^s| + |x^c - \bar{x}^c| \right) \tag{8.10}
$$

Proof. From the definition of $\tilde{g}_1$, using Lemma 8.2 and (8.8), it follows that

$$|\tilde{g}_1(x^s, x^c) - \tilde{g}_1(\bar{x}^s, \bar{x}^c)|$$
$$= |x^u - \bar{x}^u|$$
$$= |\left(\Pi^u_{m_2} D\tilde{T}(x_1)|_{X^u_{m_1}}\right)^{-1} \Big(g_2(\Pi^s_{m_2} D\tilde{T}(x_1)(x^u + x^s + x^c), \Pi^c_{m_2} D\tilde{T}(x_1)(x^u + x^s + x^c))$$
$$\qquad\qquad - g_2(\Pi^s_{m_2} D\tilde{T}(x_1)(\bar{x}^u + \bar{x}^s + \bar{x}^c), \Pi^c_{m_2} D\tilde{T}(x_1)(\bar{x}^u + \bar{x}^s + \bar{x}^c))$$
$$\qquad - \Pi^u_{m_2} D\tilde{T}(x_1)(x^s - \bar{x}^s + x^c - \bar{x}^c)\Big)|$$
$$\leq \|\left(\Pi^u_{m_2} D\tilde{T}(x_1)|_{X^u_{m_1}}\right)^{-1}\| \Big(\theta(|\Pi^s_{m_2} D\tilde{T}(x_1)(x^u - \bar{x}^u + x^s - \bar{x}^s + x^c - \bar{x}^c)|$$
$$\qquad\qquad + |\Pi^c_{m_2} D\tilde{T}(x_1)(x^u - \bar{x}^u + x^s - \bar{x}^s + x^c - \bar{x}^c)|)$$
$$\qquad\qquad + |\Pi^u_{m_2} D\tilde{T}(x_1)(x^s - \bar{x}^s + x^c - \bar{x}^c)|\Big)$$
$$\leq (C\sigma + O(\epsilon))|x^u - \bar{x}^u| + (C\sigma + O(\epsilon))(|x^s - \bar{x}^s| + |x^c - \bar{x}^c|)$$
$$\qquad + \|\left(\Pi^u_{m_2} D\tilde{T}(x_1)|_{X^u_{m_1}}\right)^{-1}\| \|\theta\Big(\|\Pi^s_{m_2} D\tilde{T}(x_1)|_{X^s_{m_1}}\| |x^s - \bar{x}^s| + \|\Pi^c_{m_2} D\tilde{T}|_{X^c_{m_1}}\| |x^c - \bar{x}^c|\Big)$$
$$\leq (C\sigma + O(\epsilon))|x^u - \bar{x}^u| + (\lambda\theta + C\sigma + O(\epsilon))(|x^s - \bar{x}^s| + |x^c - \bar{x}^c|)$$

which implies

$$|x^u - \bar{x}^u|$$
$$\leq \frac{\lambda\theta + C\sigma + O(\epsilon)}{1 - (C\sigma + O(\epsilon))} (|x^s - \bar{x}^s| + |x^c - \bar{x}^c|)$$
$$\leq \theta(|x^s - \bar{x}^s| + |x^c - \bar{x}^c|) \tag{8.11}$$

by choosing ϵ^* and σ sufficiently small. The proof is complete. $\square$

Next we want to show that the jet equivalence class $[\tilde{g}_1]$ of $\tilde{g}_1$ does not depend on the choice of $g_2 \in j_2$. Let f_2 be another Lipschitz representative of j_2 satisfying

$$|f_2(x^s, x^c) - f_2(\bar{x}^s, \bar{x}^c)|$$
$$\leq \theta(|x^s - \bar{x}^s| + |x^c - \bar{x}^c|), \quad x^\alpha, \bar{x}^\alpha \in B^\alpha_2(0, \bar{r}_2), \alpha = s, c.$$

Let $\tilde{f}_1$ denote the Lipschitz function given by Lemma 8.3 and defined on $B^s_1(0, \bar{r}_1) \times B^c_1(0, \bar{r}_1)$.

Lemma 8.5. $[\tilde{g}_1] = [\tilde{f}_1]$.

Proof. Let $(x^s, x^c) \in B^s_1(0, \hat{r}_1) \times B^c_1(0, \hat{r}_1)$, where $\hat{r}_1 = \min\{r_1, \bar{r}_1\}$. Set

$$x^u = \tilde{g}_1(x^s, x^c), \quad \bar{x}^u = \tilde{f}_1(x^s, x^c).$$

We want to show

$$\|[\tilde{g}_1] - [\tilde{f}_1]\| = \lim_{(x^s, x^c) \to (0,0)} \frac{|\tilde{g}_1(x^s, x^c) - \tilde{f}_1(x^s, x^c)|}{|x^s| + |x^c|} = 0.$$

Note that from $[g_2] = [f_2]$, $|g_2(x^s, x^c) - f_2(x^s, x^c)| = o(|x^s| + |x^c|)$ as $(x^s, x^c) \to 0$. From the definition of $\tilde{g}_1$ and $\tilde{f}_1$ and (8.8), we obtain

$$|x^u - \tilde{x}^u|$$

$$= \left| \left(\Pi^u_{m_2} D\tilde{T}(x_1)|_{X^u_{m_1}} \right)^{-1} \left(g_2(\Pi^s_{m_2} D\tilde{T}(x_1)(x^u + x^s + x^c), \Pi^c_{m_2} D\tilde{T}(x_1)(x^u + x^s + x^c)) \right. \right.$$

$$\left. \left. - f_2(\Pi^s_{m_2} D\tilde{T}(x_1)(\tilde{x}^u + x^s + x^c), \Pi^c_{m_2} D\tilde{T}(x_1)(\tilde{x}^u + x^s + x^c)) \right) \right|$$

$$\leq (C\sigma - O(\epsilon))|x^u - \tilde{x}^u| + o\left(|\Pi^s_{m_2} D\tilde{T}(x_1)(\tilde{x}^u + x^s + x^c)| + |\Pi^c_{m_2} D\tilde{T}(x_1)(\tilde{x}^u + x^s + x^c)| \right)$$

which implies together with $|\tilde{x}^u| \leq \theta(|x^s| + |x^c|)$

$$|x^u - \tilde{x}^u| = o(|x^s| + |x^c|).$$

The proof is complete. $\qquad\qquad\qquad\qquad\qquad\qquad\qquad\qquad\qquad\qquad\qquad\qquad$ □

We observe that for the given g_2, the function $\tilde{g}_1$ satisfying (8.4) is locally unique, even within the class of continuous functions. We record it here as a lemma but the proof follows from the definition of E and the uniqueness of the fixed point.

Lemma 8.6. *Let $g_1 : B^s_1(0, \bar{r}_1) \times B^s_1(0, \bar{r}_1) \to X^u_{m_1}$ be a continuous map satisfying $g_1(0,0) = 0$ and for $(x^s, x^c) \in B^s_1(0, \bar{r}_1) \times B^s_1(0, \bar{r}_1)$ there exists $(\bar{x}^s, \bar{x}^c) \in B^s_2(0, r_2) \times B^s_2(0, r_2)$ such that*

$$D\tilde{T}(x_1)(x^s + x^c + g_1(x^s, x^c)) = \bar{x}^s + \bar{x}^c + g_2(\bar{x}^s, \bar{x}^c). \tag{8.12}$$

Then $g_1(x^s, x^c) = \tilde{g}_1(x^s, x^c)$ on $B^s_1(0, \hat{r}_1) \times B^c_1(0, \hat{r}_1)$ for some $\hat{r}_1 > 0$, where $\tilde{g}_1$ is obtained from Lemma 8.4.

Thus, for each given $\gamma \in \Sigma^{cs,l}_\theta$ one may define $\tilde{\gamma} \in \Sigma^{cs,l}_\theta$ by

$$\tilde{\gamma}(m_1 + x^u_1 + x^s_1) = [\tilde{g}_1]. \tag{8.13}$$

The following summarizes what we have so far:

Proposition 8.7. *There exist $\epsilon^* > 0$ and $\sigma > 0$ such that if $\epsilon < \epsilon^*$ and $\|\tilde{T} - T^t\|_1 < \sigma$, then for each $\gamma \in \Sigma^{cs,l}_\theta$ there exists a unique $\tilde{\gamma} \in \Sigma^{cs,l}_\theta$ such that $D\tilde{T}$ maps $\tilde{\gamma}$ to γ in the sense that when $x_1 \in \tilde{W}^{cs}(\epsilon)$ and $x_2 = \tilde{T}(x_1)$, for any Lipschitz representative g_2 of $\gamma(x_2)$, there exists a Lipschitz representative $\tilde{g}_1$ of $\tilde{\gamma}(x_1)$ such that (8.4) holds, i.e.,*

$$D\tilde{T}(x_1)(x^s + x^c + \tilde{g}_1(x^s, x^c)) = \bar{x}^s + \bar{x}^c + g_2(\bar{x}^s, \bar{x}^c)$$

locally in the coordinates based at m_1 and m_2, respectively.

Define

$$\mathcal{F}(\gamma) = \tilde{\gamma}.$$

We claim

Lemma 8.8. $\mathcal{F}$ *is a contraction on* $\Sigma_\theta^{cs,l}$.

Proof. Let $\gamma, \gamma^* \in \Sigma_\theta^{cs,l}$. For $m_1 + x_1^u + x_1^s \in \tilde{W}^{cs}(\epsilon)$, we again write

$$\tilde{T}(m_1 + x_1^u + x_1^s) = m_2 + x_2^u + x_2^s$$

and $\bar{m} = T^t(m_1)$. From Proposition 8.7, there exist Lipschitz functions $\tilde{g}_1, \tilde{g}_1^* :$ $B_1^s(0, r_1) \times B_1^c(0, r_1) \to B_1^u(0, r_1)$ such that (8.10) holds and

$$[\tilde{g}_1] = \mathcal{F}(\gamma)(m_1 + x_1^u + x_1^s),$$
$$[\tilde{g}_1^*] = \mathcal{F}(\gamma^*)(m_1 + x_1^u + x_1^s).$$

Let $x^u = \tilde{g}_1(x^s, x^c)$ and $\bar{x}^u = \tilde{g}_1^*(x^s, x^c)$. From the definition of $\tilde{g}_1$ and $\tilde{g}_1^*$

$$\tilde{g}_1(x^s, x^c) - \tilde{g}_1^*(x^s, x^c)$$
$$= x^u - \bar{x}^u$$
$$= \left(\Pi_{m_2}^u D\tilde{T}(x_1)|_{X_{m_1}^u} \right)^{-1} \Big(g_2(\Pi_{m_2}^s D\tilde{T}(x_1)(x^u + x^s + x^c), \Pi_{m_2}^c D\tilde{T}(x_1)(x^u + x^s + x^c))$$
$$- g_2^*(\Pi_{m_2}^s D\tilde{T}(x_1)(\bar{x}^u + x^s + x^c), \Pi_{m_2}^c D\tilde{T}(x_1)(\bar{x}^u + x^s + x^c)) \Big)$$

where g_2 and g_2^* are Lipschitz representatives of γ and γ^*, respectively, at $m_2 + x_2^u + x_2^s$ which satisfy (8.3). Using (8.8),(8.10) and Lemma 8.2 we obtain

$$|x^u - \bar{x}^u|$$
$$\leq \| \left(\Pi_{m_2}^u D\tilde{T}(x_1)|_{X_{m_1}^u} \right)^{-1} \| \Big(\theta(|\Pi_{m_2}^s D\tilde{T}(x_1)(x^u - \bar{x}^u)| + |\Pi_{m_2}^c D\tilde{T}(x_1)(x^u - \bar{x}^u)|)$$
$$+ \|[g_2] - [g_2^*]\|(|\Pi_{m_2}^s D\tilde{T}(x_1)(\bar{x}^u + x^s + x^c)| + |\Pi_{m_2}^c D\tilde{T}(x_1)(\bar{x}^u + x^s + x^c)|)$$
$$+ o\Big(|\Pi_{m_2}^s D\tilde{T}(x_1)(\bar{x}^u + x^s + x^c)| + |\Pi_{m_2}^c D\tilde{T}(x_1)(\bar{x}^u + x^s + x^c)|\Big)$$
$$\leq (C\sigma + O(\epsilon))|x^u - \bar{x}^u| + (\lambda + C\sigma + O(\epsilon))\|[g_2] - [g_2^*]\|(|x^s| + |x^c|) + o(|x^s| + |x^c|),$$

here $|\bar{x}^u| \leq \theta(|x^s| + |x^c|)$ is used. Therefore,

$$|x^u - \bar{x}^u| \leq \frac{\lambda + C\sigma + O(\epsilon)}{1 - (C\sigma + O(\epsilon))}\|[g_2] - [g_2^*]\|(|x^s| + |x^c|) + o(|x^s| + |x^c|),$$

which yields

$$\|[\tilde{g}_1] - [\tilde{g}_1^*]\| \leq \frac{\lambda + C\sigma + O(\epsilon)}{1 - (C\sigma + O(\epsilon))}\|[g_2] - [g_2^*]\|.$$

Since $0 < \lambda < 1$, for $\lambda_1 \in (\lambda, 1)$ we may choose ϵ^* and σ so small that

$$\frac{\lambda + C\sigma + O(\epsilon)}{1 - (C\sigma + O(\epsilon))} < \lambda_1.$$

Thus

$$\|\mathcal{F}(\gamma) - \mathcal{F}(\gamma^*)\| \leq \lambda_1 \|\gamma - \gamma^*\|. \tag{8.14}$$

This completes the proof. $\qquad\qquad\qquad\qquad\qquad\qquad\qquad\qquad\qquad\qquad\qquad$ $\square$

Our goal is to find a unique fixed point of $\mathcal{F}$ in $\Sigma_\theta^{cs,l}$. The difficulty here is that $\Sigma_\theta^{cs,l}$ may not be a complete space. On the other hand, from (8.14), we have

$$\|\mathcal{F}^{(k)}(\gamma) - \mathcal{F}^{(k)}(\gamma^*)\| \leq \lambda_1^k \|\gamma - \gamma^*\|,$$

which yields that at each $m_1 + x_1^u + x_1^s \in \tilde{W}^{cs}(\epsilon)$

$$\mathcal{F}^k(\gamma)(m_1 + x_1^u + x_1^s)$$

is a Cauchy sequence in $J^c(m_1)$. Since $J^c(m_1)$ is a Banach space, we have

$$\mathcal{F}^k(\gamma)(m_1 + x_1^u + x_1^s) \rightarrow \gamma_0(m_1 + x_1^u + x_1^s),$$

where $\gamma_0(m_1 + x_1^u + x_1^s) \in J^c(X_{m_1}^s \times X_{m_1}^c, X_{m_1}^u; 0, 0)$. Clearly, the limit γ_0 is unique and does not depend on the initial choice of γ. We shall show that $\gamma_0 \in \Sigma_\theta^{cs,l}$.

Recall that the center-stable manifold $\tilde{W}^{cs}(\epsilon)$ obtained in Section 7 is given by $\tilde{W}^{cs}(\epsilon) = \Theta(\mathrm{gr}(\tilde{h}^{cs}))$, which has local coordinate representatives $\tilde{g}^{cs}$ and $\tilde{f}^{cs}$ by lemmas 7.1 and 7.2. However, $\tilde{g}^{cs}$ and $\tilde{f}^{cs}$ do not represent all points on $\tilde{W}^{cs}(\epsilon)$. From Lemma 7.1 one obtains that for all $m \in M \cap B(m_0, \rho\epsilon)$ and $x^s \in X^s((1-\eta)\rho^{-1}\epsilon)$, $\tilde{h}^{cs}(m, x^s)$ may be written as

$$\tilde{h}^{cs}(m, x^s) = \Pi_m^u \tilde{g}^{cs}(m, (\Pi_m^s|_{X_{m_0}^s})^{-1} x^s)$$

where $\tilde{g}^{cs}$ is a Lipschitz map from $(M \cap B(m_0, \rho\epsilon)) \times X_{m_0}^s(\rho^{-1}\epsilon)$ to $X_{m_0}^u$. And from Lemma 7.2 one has that for all $m \in M \cap B(m_0, \rho^{-1}\epsilon)$ and $x^s \in X_m^s(\rho^{-3}\epsilon)$

$$m + x^s + \tilde{h}^{cs}(m, x^s) = m_0 + x_0^s + x_0^c + \tilde{f}^{cs}(x_0^s, x_0^c)$$

for some $(x_0^s, x_0^c) \in X_{m_0}^s(\rho^{-2}\epsilon) \times X_{m_0}^c(\epsilon)$ and $\tilde{f}^{cs}$ is a Lipschitz map from $X_{m_0}^s(\rho^{-2}\epsilon) \times X_{m_0}^c(\epsilon)$ to $X_{m_0}^u(\epsilon)$. On the other hand from the Proposition 7.9, we have for $\epsilon < \hat{\epsilon} < \epsilon^*$

$$\tilde{W}^{cs}(\epsilon) \subset \tilde{W}^{cs}(\hat{\epsilon}),$$

thus one may choose ϵ smaller such that each point $m + x^s + \tilde{h}^{cs}(m, x^s) \in \tilde{W}^{cs}(\epsilon)$ is expressed as

$$m + x^s + \tilde{h}^{cs}(m, x^s) = m + x^s + \tilde{f}^{cs}(x^s, 0)$$

where $\tilde{f}^{cs}$ is determined by $\hat{\epsilon}$ and m, and $\tilde{h}^{cs}(m, x^s)$ is equal to $\Pi_m^u \tilde{g}^{cs}(m, (\Pi_m^s|_{X_{m_0}^s})^{-1} x^s)$.

For the sake of convenience, we shall drop the superscript cs and tilde on $\tilde{f}^{cs}$ in our proofs. Let $m_1 + x_1^u + x_1^s \in \tilde{W}^{cs}(\epsilon)$. From the above discussion, there is a Lipschitz function $f : X_{m_1}^s(\epsilon) \times X_{m_1}^c(\epsilon) \to X_{m_1}^u(\epsilon)$ such that $m_1 + x_1^u + x_1^s = m_1 + x_1^s + f(x_1^s, 0)$. Let $f_1(x^s, x^c) = f(x^s + x_1^s, x^c) - f(x_1^s, 0)$. We have $f_1(0,0) = 0$ and

$$|f_1(x^s, x^c) - f_1(\bar{x}^s, \bar{x}^c)| \le \theta(|x^s - \bar{x}^s| + |x^c - \bar{x}^c|),$$

for $(x^s, x^c), (\bar{x}^s, \bar{x}^c) \in B_1^s(0, r_1) \times B_1^c(0, r_1)$ for some $r_1 > 0$. Thus f_1 induces $\tilde{\gamma}_0 \in \Sigma_\theta^{cs,l}$ by

$$\tilde{\gamma}_0(m_1 + x_1^u + x_1^s) = [f_1].$$

Lemma 8.9. $\mathcal{F}(\tilde{\gamma}_0) = \tilde{\gamma}_0$ and hence $\gamma_0 = \tilde{\gamma}_0$.

Proof. For $x_1 = m_1 + x_1^u + x_1^s \in \tilde{W}^{cs}(\epsilon)$ we write $m_2 + x_2^u + x_2^s = \tilde{T}(x_1)$ and denote the local coordinate representative of $\tilde{W}^{cs}(\epsilon)$ at $m_2 + x_2^u + x_2^s$ by

$$m_2 + \bar{x}^s + \bar{x}^c + \bar{f}(\bar{x}^s, \bar{x}^c).$$

Let

$$f_2(\bar{x}^s, \bar{x}^c) = \bar{f}(\bar{x}^s + x_2^s, \bar{x}^c) - \bar{f}(x_2^s, 0).$$

Let $\tilde{f}_1$ be given by Lemma 8.3 from f_2. We want to show $[f_1] = [\tilde{f}_1]$. For $(x^s, x^c) \in B_1^s(0, r_1) \times B_1^c(0, r_1)$ let

$$x^u = f_1(x^s, x^c)$$
$$\tilde{x}^u = \tilde{f}_1(x^s, x^c)$$

From the definition of $\tilde{f}_1$,

$$\tilde{x}^u = \left(\Pi_{m_2}^u D\tilde{T}(x_1)|_{X_{m_1}^u}\right)^{-1} \left(f_2(\Pi_{m_2}^s D\tilde{T}(x_1)(\tilde{x}^u + x^s + x^c), \Pi_{m_2}^c D\tilde{T}(x_1)(\tilde{x}^u + x^s + x^c)) - \Pi_{m_2}^u D\tilde{T}(x_1)(x^s + x^c)\right). \tag{8.15}$$

On the other hand, from the invariance of $W^{cs}(\epsilon)$, there exists $(\bar{x}^s, \bar{x}^c) \in X_{m_2}^s(\rho^{-2}\epsilon) \times X_{m_1}^c(\epsilon)$ such that

$$\tilde{T}(m_1 + x_1^u + x_1^s + x^u + x^s + x^c)$$
$$= m_2 + x_2^u + x_2^s + \bar{x}^u + \bar{x}^s + \bar{x}^c$$

where $x^u = f_1(x^s, x^c)$ and $\bar{x}^u = f_2(\bar{x}^s, \bar{x}^c)$.

By the Taylor expansion, we obtain

$$\bar{x}^u + \bar{x}^s + \bar{x}^c = D\tilde{T}(x_1)(x^u + x^s + x^c) + o(|x^u| + |x^s| + |x^c|).$$

Note that $|x^u| \le \theta(|x^s| + |x^c|)$. Hence,

$$\bar{x}^u + \bar{x}^s + \bar{x}^c = D\tilde{T}(x_1)(x^u + x^s + x^c) + o(|x^s| + |x^c|)$$

and for $\alpha = u, s, c$

$$\bar{x}^\alpha = \Pi^\alpha_{m_2} D\tilde{T}(x_1)(x^u + x^s + x^c) + o(|x^s| + |x^c|).$$

In particular,

$$\bar{x}^u = \Pi^u_{m_2} D\tilde{T}(x_1)(x^u + x^s + x^c) + o(|x^s| + |x^c|)$$
$$= \Pi^u_{m_2} D\tilde{T}(x_1)(x^u) + \Pi^u_{m_2} D\tilde{T}(x_1)(x^s + x^c) + o(|x^s| + |x^c|)$$

which yields

$$x^u = \left(\Pi^u_{m_2} D\tilde{T}(x_1)|_{X^u_{m_1}}\right)^{-1} \left(\bar{x}^u - \Pi^u_{m_2} D\tilde{T}(x_1)(x^s + x^c)\right) + o(|x^s| + |x^c|)$$
$$= \left(\Pi^u_{m_2} D\tilde{T}(x_1)|_{X^u_{m_1}}\right)^{-1} \left(f_2(\bar{x}^s, \bar{x}^c) - \Pi^u_{m_2} D\tilde{T}(x_1)(x^s + x^c)\right) + o(|x^s| + |x^c|)$$
$$= \left(\Pi^u_{m_2} D\tilde{T}(x_1)|_{X^u_{m_1}}\right)^{-1} \left(f_2(\Pi^s_{m_2} D\tilde{T}(x_1)(x^u + x^s + x^c) + o(|x^s| + |x^c|),\right.$$
$$\Pi^c_{m_2} D\tilde{T}(x_1)(x^u + x^s + x^c) + o(|x^s| + |x^c|))$$
$$\left. - \Pi^u_{m_2} D\tilde{T}(x_1)(x^s + x^c)\right) + o(|x^s| + |x^c|).$$

Using (8.5), (8.8), (8,15) and Lemma 8.2, we obtain

$$|\tilde{x}^u - x^u|$$
$$\le \| \left(\Pi^u_{m_2} D\tilde{T}(x_1)|_{X^u_{m_1}}\right)^{-1} \| \theta \left(|\Pi^s_{m_2} D\tilde{T}(x_1)(\tilde{x}^u - x^u)| + |\Pi^c_{m_2} D\tilde{T}(x_1)(\tilde{x}^u - x^u)|\right)$$
$$+ o(|x^s| + |x^c|)$$
$$\le (C\sigma + O(\epsilon))|\tilde{x}^u - x^u| + o(|x^s| + |x^c|).$$

Thus, by choosing ϵ^* and σ so small that $C\sigma + O(\epsilon) < 1/2$, we obtain

$$|\tilde{x}^u - x^u| \le o(|x^s| + |x^c|),$$

namely,

$$|f_1(x^s, x^c) - \tilde{f}_1(x^s, x^c)| \le o(|x^s| + |x^c|)$$

and hence

$$[f_1] = [\tilde{f}_1].$$

The uniqueness of the limit γ_0 implies that $\gamma_0 = \tilde{\gamma}_0$. This completes the proof. $\square$

We are now ready to show

Proposition 8.10. $f(x^s, x^c)$ *is differentiable.*

Proof. It is clear that from the definition of $\mathcal{F}$, particularly the definition of E, we have that if $\gamma \in \Sigma_\theta^{cs,a}$ then $\mathcal{F}(\gamma) \in \Sigma_\theta^{cs,a}$. Since $J^a(m) = J^d(m)$ is closed, $\gamma_0(m_1 + x_1^u + x_1^s) \in J^a(m_1)$, where γ_0 is the limit of the Cauchy sequence $\mathcal{F}^k(\gamma)$. Since $\gamma_0 = \tilde{\gamma}_0$ is unique, for $m_1 + x_1^u + x_1^s \in \tilde{W}^{cs}(\epsilon)$, we have $[f_1] \in J^a(m_1)$, that is, f is differentiable at $(x_1^s, 0)$. Next, we show $f(x^s, x^c)$ is differentiable in $X_{m_1}^s(\epsilon) \times X_{m_1}^c(\epsilon)$.

For $(x_0^s, x_0^c) \in X_{m_1}^s(\epsilon) \times X_{m_1}^c(\epsilon)$ write $m_1 + x_0^s + x_0^c + f(x_0^s, x_0^c)$ as

$$m_1 + x_0^s + x_0^c + f(x_0^s, x_0^c) = m_2 + \bar{x}_2^u + \bar{x}_2^s. \tag{8.16}$$

Any point on $\tilde{W}^{cs}(\epsilon)$ near this point may be written as

$$\begin{aligned} &m_1 + x_0^s + x_0^c + x^s + x^c + f(x_0^s + x^s, x_0^c + x^c) \\ &= m_2 + \bar{x}_2^s + x_2^s + x_2^c + \bar{f}(\bar{x}_2^s + x_2^s, x_2^c), \end{aligned} \tag{8.17}$$

where $\bar{f}$ is the representative of $\tilde{W}^{cs}(\epsilon)$ at m_2. Subtracting (8.16) from (8.17), we get

$$\begin{aligned} &x^s + x^c + f(x_0^s + x^s, x_0^c + x^c) - f(x_0^s, x_0^c) \\ &= x_2^s + x_2^c + \bar{f}(\bar{x}_2^s + x_2^s, x_2^c) - \bar{f}(\bar{x}_2^s, 0). \end{aligned} \tag{8.18}$$

To simplify computation, let

$$f_1(x^s, x^c) = f(x_0^s + x^s, x_0^c + x^c) - f(x_0^s, x_0^c),$$

and

$$f_2(x_2^s, x_2^c) = \bar{f}(\bar{x}_2^s + x_2^s, x_2^c) - \bar{f}(\bar{x}_2^s, 0).$$

Thus (8.18) has the form

$$x^s + x^c + f_1(x^s, x^c) = x_2^s + x_2^c + f_2(x_2^s, x_2^c) \tag{8.19}$$

Applying the projection $\Pi_{m_1}^\alpha$ to (8.19) for $\alpha = s, c$, we obtain

$$|x^\alpha| \le \|\Pi_{m_1}^\alpha\|(1 + \theta)(|x_2^s| + |x_2^c|).$$

Similarly, we have

$$|x_2^\alpha| \le \|\Pi_{m_2}^\alpha\|(1 + \theta)(|x^s| + |x^c|). \tag{8.20}$$

Note that f_2 is differentiable at $(0,0)$. Hence from (8.19), we obtain

$$\begin{aligned} f_1(x^s, x^c) &= \Pi_{m_1}^u(x_2^s + x_2^c) + \Pi_{m_1}^u Df_2(0,0)(x_2^s, x_2^c) + o(|x_2^s| + |x_2^c|) \\ &= \Pi_{m_1}^u \left(\Pi_{m_2}^s + \Pi_{m_2}^c \right)(x^s + x^c + f_1(x^s, x^c)) \\ &\quad + \Pi_{m_1}^u Df_2(0,0) \left(\Pi_{m_2}^s(x^s + x^c + f_1(x^s, x^c)), \Pi_{m_2}^c(x^s + x^c + f_1(x^s, x^c)) \right) \\ &\quad + o(|x^s| + |x^c|). \end{aligned}$$

Let $G \equiv I - \Pi^u_{m_1} \left(\Pi^s_{m_2} + \Pi^c_{m_2} \right) \Pi^u_{m_1} - \Pi^u_{m_1} Df_2(0,0) \left(\Pi^s_{m_2} \Pi^u_{m_1}, \Pi^c_{m_2} \Pi^u_{m_1} \right)$. Thus the above identity may be written as

$$
\begin{aligned}
Gf_1(x^s, x^c) = {}& \Pi^u_{m_1} \left(\Pi^s_{m_2} + \Pi^c_{m_2} \right) (x^s + x^c) \\
& + \Pi^u_{m_1} Df_2(0,0) \left(\Pi^s_{m_2}(x^s + x^c), \Pi^c_{m_2}(x^s + x^c) \right) \\
& + o(|x^s| + |x^c|)
\end{aligned}
\tag{8.21}
$$

Note that G may also be written as

$$
\begin{aligned}
G = {}& I - \Pi^u_{m_1} \left(\Pi^s_{m_2} - \Pi^s_{m_1} + \Pi^c_{m_2} - \Pi^c_{m_1} \right) \Pi^u_{m_1} \\
& - \Pi^u_{m_1} Df_2(0,0) \left((\Pi^s_{m_2} - \Pi^s_{m_1})\Pi^u_{m_1}, (\Pi^c_{m_2} - \Pi^c_{m_1})\Pi^u_{m_1} \right).
\end{aligned}
$$

Hence,

$$
\begin{aligned}
\|I - G\| \leq {}& \|\Pi^u_{m_1}\|^2 \left(\|\Pi^s_{m_2} - \Pi^s_{m_1}\| + \|\Pi^c_{m_2} - \Pi^c_{m_1}\| \right) \\
& + \|\Pi^u_{m_1}\|^2 \theta \left(\|\Pi^s_{m_2} - \Pi^s_{m_1}\| + \|\Pi^c_{m_2} - \Pi^c_{m_1}\| \right) \\
= {}& (1 + \theta)\|\Pi^u_{m_1}\|^2 \left(\|\Pi^s_{m_2} - \Pi^s_{m_1}\| + \|\Pi^c_{m_2} - \Pi^c_{m_1}\| \right).
\end{aligned}
$$

Note that from (8.16) $|m_2 - m_1| < 5\epsilon$. Thus, from (H4) there exists $\epsilon^* > 0$ such that if $\epsilon < \epsilon^*$, then

$$
(1 + \theta)\|\Pi^u_{m_1}\|^2 \left(\|\Pi^s_{m_2} - \Pi^s_{m_1}\| + \|\Pi^c_{m_2} - \Pi^c_{m_1}\| \right) < \frac{1}{2},
$$

which yields $\|I - G\| \leq \frac{1}{2}$ and that G is an isomorphism from $X^u_{m_1}$ onto $X^u_{m_1}$. Denote the inverse by G^{-1}, which is a bounded linear operator. From (8.21), we obtain

$$
f_1(x^s, x^c) = G^{-1} R(x^s + x^c) + o(|x^s| + |x^c|),
$$

where $R = \Pi^u_{m_1} \left(\Pi^s_{m_2} + \Pi^c_{m_2} \right) - \Pi^u_{m_1} Df_2(0,0) \left(\Pi^s_{m_2} \Pi^u_{m_1}, \Pi^c_{m_2} \Pi^u_{m_1} \right)$ is a bounded linear operator from $X^u_{m_1}$ into itself. Therefore f_1 is differentiable at $(x^s, x^c) = (0,0)$. The proof is complete. □

Finally, we want to show Df is C^0.

Let $L(X^s \oplus X^c, X^u)$ denote the vector bundle over M with fiber $L(X^s_m \oplus X^c_m, X^u_m)$, where $L(X^s_m \oplus X^c_m, X^u_m)$ is a Banach space of all bounded linear operators from $X^s_m \oplus X^c_m$ to X^u_m. For each $m_0 \in M$, let $\eta < \sqrt{2} - 1$ and U be an η-neighborhood of m_0. Thus, from (4.5) we may define a trivialization

$$
\tilde{\Psi}^{cs}_{m_0} : U \cap M \times L(X^s_{m_0} \oplus X^c_{m_0}, X^u_{m_0}) \rightarrow L(X^s \oplus X^c, X^u)
$$

by

$$
\tilde{\Psi}^{cs}_{m_0}(m, L(m_0)) = \Pi^u_m L(m_0) \left((\Pi^s_m |_{X^s_{m_0}})^{-1} \oplus (\Pi^c_m |_{X^c_{m_0}})^{-1} \right).
\tag{8.22}
$$

It is not hard to see that $L(X^s \oplus X^c, X^u)$ is a Finsler bundle with the trivialization $\tilde{\Psi}^{cs}_{m_0}$ which gives an isomorphism from $L(X^s_{m_0} \oplus X^c_{m_0}, X^u_{m_0})$ onto $L(X^s_m \oplus X^c_m, X^u_m)$, the inverse of which is $\tilde{L} \to (\Pi^u_m|_{X^u_{m_0}})^{-1} \tilde{L}(\Pi^s_m \Pi^s_{m_0} + \Pi^c_m \Pi^c_{m_0})$.

Define the space

$$\Lambda^{cs}_\theta = \Big\{ \ell : \tilde{W}^{cs}(\epsilon) \to L(X^s \oplus X^c, X^u) \text{ is a } C^0 \text{ section and}$$

$$\|\ell\| \leq \theta \}$$

with the norm

$$\|\ell\| = \sup \Big\{ \|\ell(x)\| \ : \ x \in \tilde{W}^{cs}(\epsilon) \Big\}.$$

Note that $\|\ell(x)\| = sup\{|\ell(x)(\tilde{x}^s, \tilde{x}^c)| : \tilde{x}^\alpha \in X^\alpha_m, \alpha = s, c, |\tilde{x}^s| + |\tilde{x}^c| \leq 1\}$. It is easy to check Λ^{cs}_θ is a Banach space. The space Λ^{cs}_θ may be regarded as a subset of $\Sigma^{cs,l}_\theta$ by identifying ℓ and $[\ell]$.

We shall show that $\mathcal{F}(\Lambda^{cs}_\theta) \subset \Lambda^{cs}_\theta$.

Let $x_1 = m_1 + x^u_1 + x^s_1 \in \tilde{W}^{cs}(\epsilon)$ be fixed. Consider a point $\bar{x}_1 = \bar{m}_1 + \bar{x}^u_1 + \bar{x}^s_1 \in \tilde{W}^{cs}(\epsilon)$ sufficiently close to x_1. We write

$$\tilde{T}(x_1) = m_2 + x^u_2 + x^s_2 = x_2$$

and

$$\tilde{T}(\bar{x}_1) = \bar{m}_2 + \bar{x}^u_2 + \bar{x}^s_2 = \bar{x}_2.$$

Let $B(x_1, r)$ be a ball in X. We may choose r small enough that if $\bar{x}_1 \in B(x_1, r) \cap \tilde{W}^{cs}(\epsilon)$, then $\bar{m}_1$ is in an η-neighborhood of m_1 and $\bar{m}_2$ is in an η-neighborhood of m_2.

Let $\ell \in \Lambda^{cs}_\theta$ and denote $\mathcal{F}(\ell)$ by $\tilde{\ell}$. To show $\tilde{\ell} \in \Lambda^{cs}_\theta$, it is enough to show

Proposition 8.11. *There exist $\epsilon^* > 0$ and $\sigma > 0$ such that if $\epsilon < \epsilon^*$ and $\|\tilde{T} - T^t\| < \sigma$, then*

$$\lim_{\bar{x}_1 \to x_1} \|\tilde{\ell}(x_1) - \Big(\Pi^u_{\bar{m}_1}|_{X^u_{m_1}}\Big)^{-1} \tilde{\ell}(\bar{x}_1) \Big(\Pi^s_{\bar{m}_1} \Pi^s_{m_1} + \Pi^c_{\bar{m}_1} \Pi^c_{m_1}\Big)\| = 0.$$

In order to prove this proposition, we need some lemmas. Let

$$S(x_1) = \Big(\Pi^u_{m_2} D\tilde{T}(x_1)|_{X^u_{m_1}}\Big)^{-1} \Pi^u_{m_2} \tag{8.23}$$

$$\bar{S}(\bar{x}_1) = \Big(\Pi^u_{\bar{m}_1}|_{X^u_{m_1}}\Big)^{-1} \Big(\Pi^u_{\bar{m}_2} D\tilde{T}(\bar{x}_1)|_{X^u_{m_1}}\Big)^{-1} \Pi^u_{\bar{m}_2}. \tag{8.24}$$

From Lemma 8.2, S and $\bar{S}$ are well-defined. Furthermore,

Lemma 8.12.

$$\lim_{\bar{x}_1 \to x_1} \|\bar{S}(\bar{x}_1) - S(x_1)\| = 0$$

Proof. We first notice that from the continuity of $D\tilde{T}$ and Lemma 4.3 it follows that $\bar{x}_2 \to x_2, \bar{m}_1 \to m_1$ and $\bar{m}_2 \to m_2$ as $\bar{x}_1 \to x_1$. We consider only $\bar{x}_1 \in B(x_1, r)$. Thus $\bar{m}_1$ and $\bar{m}_2$ are in an η-neighborhoods of m_1 and m_2, respectively. For $x \in X$, let $y = S(x_1)x$ and $z = \bar{S}(\bar{x}_1)x$. Note that $y, z \in X^u_{m_1}$. Hence,

$$\Pi^u_{m_2} D\tilde{T}(x_1)y = \Pi^u_{m_2}x \tag{8.25}$$

$$\Pi^u_{\bar{m}_2} D\tilde{T}(\bar{x}_1)\Pi^u_{\bar{m}_1} z = \Pi^u_{\bar{m}_2}x. \tag{8.26}$$

Subtracting (8.26) from (8.25), we obtain

$$\Pi^u_{m_2} D\tilde{T}(x_1)(y - z)$$
$$+ \left(\Pi^u_{m_2} D\tilde{T}(x_1)\Pi^u_{m_1} - \Pi^u_{\bar{m}_2} D\tilde{T}(\bar{x}_1)\Pi^u_{\bar{m}_1} \right) z$$
$$= \left(\Pi^u_{m_2} - \Pi^u_{\bar{m}_2} \right) x. \tag{8.27}$$

Using Lemma 8.2 and (4.5), we have from (8.26)

$$|z| \le \lambda \| \left(\Pi^u_{\bar{m}_1} |_{X^u_{\bar{m}_1}} \right)^{-1} \| \| \Pi^u_{\bar{m}_2} x | \le \frac{\lambda(1 + \eta)}{(1 - \eta)}|x|. \tag{8.28}$$

Also from (8.27)

$$|y - z|$$
$$\le \lambda \| \Pi^u_{m_2} D\tilde{T}(x_1)\Pi^u_{m_1} - \Pi^u_{\bar{m}_2} D\tilde{T}(\bar{x}_1)\Pi^u_{\bar{m}_1} \| |z| + \lambda \| \Pi^u_{m_2} - \Pi^u_{\bar{m}_2} \| |x|,$$

which yields

$$\|S(x_1) - \bar{S}(\bar{x}_1)\|$$
$$\le \lambda^2 \frac{1 + \eta}{1 - \eta} \| \Pi^u_{m_2} D\tilde{T}(x_1)\Pi^u_{m_1} - \Pi^u_{\bar{m}_2} D\tilde{T}(\bar{x}_1)\Pi^u_{\bar{m}_1} \| + \lambda \| \Pi^u_{m_2} - \Pi^u_{\bar{m}_2} \| \tag{8.29}$$

Using the continuity of Π^u_m and $D\tilde{T}$ and letting $\bar{x}_1 \to x_1$ in (8.29), we complete the proof. □

For $x^s + x^c \in X^s_{m_1} \oplus X^c_{m_1}$ with $|x^s| + |x^c| \le 1$ and let

$$x^u = \tilde{\ell}(x_1)(x^s + x^c)$$
$$\bar{x}^u = \tilde{\ell}(\bar{x}_1) \left(\Pi^s_{\bar{m}_1} x^s + \Pi^c_{\bar{m}_1} x^c \right).$$

Proof of Proposition 8.11. Since $\bar{m}_1$ is in an η-neighborhood of m_1, by (4.5), $\Pi^u_{\bar{m}_1}|_{X^u_{m_1}}$ is invertible. Thus there exists $\tilde{x}^u \in X^u_{m_1}$ such that $\bar{x}^u = \Pi^u_{\bar{m}_1}\big|_{X^u_{m_1}} \tilde{x}^u$. By the definition of $\tilde{\ell}$,

$$x^u = \left(\Pi^u_{m_2} D\tilde{T}(x_1)|_{X^u_{m_1}}\right)^{-1} \Big(\ell(x_2)((\Pi^s_{m_2} + \Pi^c_{m_2})D\tilde{T}(x_1)(x^u + x^s + x^c)) \\ - \Pi^u_{m_2}D\tilde{T}(x_1)(x^s + x^c)\Big) \tag{8.30}$$

and

$$\tilde{x}^u = \left(\Pi^u_{\bar{m}_1}|_{X^u_{m_1}}\right)^{-1}\left(\Pi^u_{\bar{m}_2}D\tilde{T}(\bar{x}_1)|_{X^u_{m_1}}\right)^{-1} \\ \Big(\ell(\bar{x}_2)((\Pi^s_{\bar{m}_2} + \Pi^c_{\bar{m}_2})D\tilde{T}(\bar{x}_1)\Big(\Pi^u_{\bar{m}_1}\tilde{x}^u + \Pi^s_{\bar{m}_1}x^s + \Pi^c_{\bar{m}_1}x^c)) \\ - \Pi^u_{\bar{m}_2}D\tilde{T}(\bar{x}_1)\Big(\Pi^s_{\bar{m}_1}x^s + \Pi^c_{\bar{m}_1}x^c\Big)\Big).$$

Because $\ell \in \Lambda^{cs}_\theta$, we have
$$|x^u| \le \theta\left(|x^s| + |x^c|\right)$$

and using (4.5),
$$|\tilde{x}^u| \le \frac{\theta(1+\eta)}{1-\eta}\left(|x^s| + |x^c|\right).$$

Next we want to estimate $|x^u - \tilde{x}^u|$. Let

$$I_1 = S(x_1)\Big(\ell(x_2)(\Pi^s_{m_2} + \Pi^c_{m_2}) - \left(\Pi^u_{\bar{m}_2}|_{X^u_{m_2}}\right)^{-1}\ell(\bar{x}_2)(\Pi^s_{\bar{m}_2}\Pi^s_{m_2} + \Pi^c_{\bar{m}_2}\Pi^c_{m_2})\Big) \\ D\tilde{T}(x_1)(x^u + x^s + x^c)$$

$$I_2 = (S(x_1) - \bar{S}(\bar{x}_1))\left(\Pi^u_{\bar{m}_1}|_{X^u_{m_2}}\right)^{-1}\ell(\bar{x}_2)\Big(\Pi^s_{\bar{m}_2}\Pi^s_{m_2} + \Pi^c_{\bar{m}_2}\Pi^c_{m_2}\Big)D\tilde{T}(x_1)(x^u + x^s + x^c)$$

$$I_3 = \bar{S}(\bar{x}_1)\ell(\bar{x}_2)\Big((\Pi^s_{\bar{m}_2}\Pi^s_{m_2} + \Pi^c_{\bar{m}_2}\Pi^c_{m_2}) \\ D\tilde{T}(x_1)(x^u - \Pi^u_{\bar{m}_1}\tilde{x}^u + x^s - \Pi^s_{\bar{m}_1}x^s + x^c - \Pi^c_{\bar{m}_1}x^c)\Big)$$

$$I_4 = \bar{S}(\bar{x}_1)\ell(\bar{x}_2)\Big(\Pi^s_{m_2}(\Pi^s_{\bar{m}_2}D\tilde{T}(x_1) - \Pi^s_{\bar{m}_2}D\tilde{T}(\bar{x}_1))(\Pi^u_{\bar{m}_1}\tilde{x}^u + \Pi^s_{\bar{m}_1}x^s + \Pi^c_{\bar{m}_1}x^c)) \\ + \Pi^c_{\bar{m}_2}(\Pi^c_{m_2}D\tilde{T}(x_1) - \Pi^c_{\bar{m}_2}D\tilde{T}(\bar{x}_1))(\Pi^u_{\bar{m}_1}\tilde{x}^u + \Pi^s_{\bar{m}_1}x^s + \Pi^c_{\bar{m}_1}x^c)\Big)$$

$$I_5 = \bar{S}(\bar{x}_1)\Pi^u_{\bar{m}_2}D\tilde{T}(\bar{x}_1)\Big(\Pi^s_{\bar{m}_1}x^s + \Pi^c\bar{m}_1x^c\Big) - S(x_1)\Pi^u_{m_2}D\tilde{T}(x_1)(x^s + x^c).$$

It is straightforward to check that

$$x^u - \tilde{x}^u = I_1 + I_2 + I_3 + I_4 + I_5. \tag{8.31}$$

From the continuity of ℓ, we have that $I_1 \to 0$ as $\bar{x}_1 \to x_1$. Lemma 8.12 yields that $I_2, I_5 \to 0$ as $\bar{x}_1 \to x_1$. The continuity of the projections and $D\tilde{T}$ gives that $I_4 \to 0$ as $\bar{x}_1 \to x_1$. Using (8.8), we may choose ϵ^* and σ sufficiently small such than

$$|I_3| \leq \frac{1}{2}|x^u - \tilde{x}^u| + C\Big(\|\Pi^u_{m_1} - \Pi^u_{\bar{m}_1}\| + \|\Pi^s_{m_1} - \Pi^s_{\bar{m}_1}\| + \|\Pi^c_{m_1} - \Pi^c_{\bar{m}_1}\|\Big).$$

Therefore, from (8.31) we have $|x^u - \tilde{x}^u| \to 0$ as $\bar{x}_1 \to x_1$. This completes the proof. $\square$

Proposition 8.13. *There exists positive constant ϵ^* such that for $\epsilon < \epsilon^*$, $Df(x^s, x^c)$ is continuous in (x^s, x^c).*

Proof. As we mentioned before, Proposition 8.11 yields that $\mathcal{F}(\Lambda^{cs}_\theta) \subset \Lambda^{cs}_\theta$. Since $\mathcal{F}$ is a contraction and Λ^{cs}_θ is complete $\mathcal{F}$ has a unique fixed point $\tilde{\ell} \in \Lambda^{cs}_\theta$. On the other hand, $\tilde{\gamma}_0$ is the unique fixed point of $\mathcal{F}$ in $\Sigma^{cs,l}_\theta$. Therefore, $\tilde{\gamma}_0 = [\tilde{\ell}]$ which yields

$$m_1 + x_1^u + x_1^s \to Df(x_1^s, 0)$$

is continuous, that is,

$$\lim_{\bar{x}_1 \to x_1} \|Df(x_1^s, 0) - \Big(\Pi^u_{\bar{m}_1}\big|_{X^u_{m_1}}\Big)^{-1} D\bar{f}(\bar{x}_1^s, 0)\begin{pmatrix} \Pi^s_{\bar{m}_1} & 0 \\ 0 & \Pi^c_{\bar{m}_1} \end{pmatrix}\| = 0, \qquad (8.32)$$

where $x_1 = m_1 + x_1^u + x_1^s, \bar{x}_1 = \bar{m}_1 + \bar{x}_1^u + \bar{x}_1^s \in \tilde{W}^{cs}(\epsilon)$.

We shall show that $Df(x^s, x^c)$ is continuous at (x_0^s, x_0^c). We write

$$m_1 + x_0^s + x_0^c + f(x_0^s, x_0^c) = m_2 + x_2^s + x_2^u \qquad (8.33)$$

and let f_2 be the representative at m_2, so $x_2^u = f_2(x_2^s, 0)$. Any point in $\tilde{W}^{cs}(\epsilon)$ nearby may uniquely be expressed as

$$m_1 + x_0^s + x_0^c + x^s + x^c + f(x_0^s + x^s, x_0^c + x^c)$$
$$= m_2 + x_2^s + \bar{x}_2^s + \bar{x}_2^c + f_2(x_2^s + \bar{x}_2^s, \bar{x}_2^c). \qquad (8.34)$$

Subtracting (8.33) from (8.34) and letting

$$\tilde{f}(x^s, x^c) = f(x_0^s + x^s, x_0^c + x^c) - f(x_0^s, x_0^c)$$

and

$$\tilde{f}_2(\bar{x}_2^s, \bar{x}_2^c) = f_2(x_2^s + \bar{x}_2^s, \bar{x}_2^c) - f_2(x_2^s, 0)$$

we obtain

$$x^s + x^c + \tilde{f}(x^s, x^c) = \bar{x}_2^s + \bar{x}_2^c + \tilde{f}_2(\bar{x}_2^s, \bar{x}_2^c), \qquad (8.35)$$

which gives

$$(x^s, x^c) = (\Pi^s_{m_1}, \Pi^c_{m_1})(\bar{x}^s_2 + \bar{x}^c_2 + \tilde{f}_2(\bar{x}^s_2, \bar{x}^c_2)) \tag{8.36}$$

and

$$\tilde{f}(x^s, x^c) = \Pi^u_{m_1}(\bar{x}^s_2 + \bar{x}^c_2 + \tilde{f}_2(\bar{x}^s_2, \bar{x}^c_2)). \tag{8.37}$$

Let

$$G_2(\bar{x}^s_2, \bar{x}^c_2) = (\Pi^s_{m_1}, \Pi^c_{m_1})(\bar{x}^s_2 + \bar{x}^c_2 + \tilde{f}_2(\bar{x}^s_2, \bar{x}^c_2)).$$

Differentiating (8.37) with respect to $(\bar{x}^s_2, \bar{x}^c_2)$ and evaluating at $(\bar{x}^s_2, \bar{x}^c_2) = (0, 0)$, we have

$$Df(x^s_0, x^c_0)DG_2(0, 0) = \Pi^u_{m_1} \oplus \Pi^u_{m_1} + \Pi^u_{m_1}Df_2(x^s_2, 0),$$

and

$$DG_2(0,0)(\bar{x}^s_2, \bar{x}^c_2) = \Big(\Pi^s_{m_1}(\bar{x}^s_2 + \bar{x}^c_2) + \Pi^s_{m_1}Df_2(x^s_2, 0)(\bar{x}^s_2, \bar{x}^c_2),$$
$$\Pi^c_{m_1}(\bar{x}^s_2 + \bar{x}^c_2) + \Pi^c_{m_1}Df_2(x^s_2, 0)(\bar{x}^s_2, \bar{x}^c_2)\Big).$$

Let $(\tilde{x}^s_0, \tilde{x}^c_0)$ be a point near (x^s_0, x^c_0). Then we may write

$$m_1 + \tilde{x}^s_0 + \tilde{x}^c_0 + f(\tilde{x}^s_0, \tilde{x}^c_0) = m_3 + x^s_3 + f_3(x^s_3, 0).$$

Similarly, we have

$$Df(\tilde{x}^s_0, \tilde{x}^c_0)DG_3(0, 0) = \Pi^u_{m_1} \oplus \Pi^u_{m_1} + \Pi^u_{m_1}Df_3(x^s_3, 0)$$

and

$$DG_3(0,0)(\bar{x}^s_3, \bar{x}^c_3) = \Big(\Pi^s_{m_1}(\bar{x}^s_3 + \bar{x}^c_3) + \Pi^s_{m_1}Df_3(x^s_3, 0)(\bar{x}^s_3, \bar{x}^c_3),$$
$$\Pi^c_{m_1}(\bar{x}^s_3 + \bar{x}^c_3) + \Pi^c_{m_1}Df_3(x^s_3, 0)(\bar{x}^s_3, \bar{x}^c_3)\Big).$$

In order to show

$$\lim_{(\tilde{x}^s_0, \tilde{x}^c_0) \to (x^s_0, x^c_0)} \|Df(\tilde{x}^s_0, \tilde{x}^c_0) - Df(x^s_0, x^c_0)\| = 0, \tag{8.38}$$

we let

$$I_1 = (Df(x^s_0, x^c_0) - Df(\tilde{x}^s_0, \tilde{x}^c_0))\,DG_2(0, 0)$$
$$I_2 = Df(\tilde{x}^s_0, \tilde{x}^c_0)\left(DG_3(0, 0)\begin{pmatrix} \Pi^s_{m_3} & 0 \\ 0 & \Pi^c_{m_3} \end{pmatrix} - DG_2(0, 0)\right)$$
$$I_3 = \Pi^u_{m_1}\left(\Pi^s_{m_2} - \Pi^s_{m_3}\right) \oplus \Pi^u_{m_1}\left(\Pi^c_{m_2} - \Pi^c_{m_3}\right)$$
$$I_4 = \Pi^u_{m_1}\left(Df_2(x^s_2, 0) - Df_3(x^s_3, 0)\begin{pmatrix} \Pi^s_{m_3} & 0 \\ 0 & \Pi^c_{m_3} \end{pmatrix}\right).$$

It is easy to verify that

$$I_1 = I_2 + I_3 + I_4. \tag{8.39}$$

By Lemma 4.3, we have that as $(\tilde{x}_0^s, \tilde{x}_0^c) \to (x_0^s, x_0^c)$

$$m_3 + x_3^s + f_3(x_3^s, 0) \to m_2 + x_2^s + f_2(x_2^s, 0)$$

and $m_3 \to m_2$. Therefore, from the continuity of Π_m^α in m, we have that

$$\lim_{(\tilde{x}_0^s, \tilde{x}_0^c) \to (x_0^s, x_0^c)} I_3 = 0.$$

Express I_4 as

$$I_4 = \Pi_{m_1}^u \left(Df_2(x_2^s, 0) - \left(\Pi_{m_3}^u | X_{m_2}^u \right)^{-1} Df_3(x_3^s, 0) \begin{pmatrix} \Pi_{m_3}^s & 0 \\ 0 & \Pi_{m_3}^c \end{pmatrix} \right)$$

$$+ \Pi_{m_1}^u (\Pi_{m_2}^u - \Pi_{m_3}^u) \left(\Pi_{m_3}^u | X_{m_2}^u \right)^{-1} Df_3(x_3^s, 0) \begin{pmatrix} \Pi_{m_3}^s & 0 \\ 0 & \Pi_{m_3}^c \end{pmatrix}.$$

Using (8.32) and the continuity of Π_m^u in m, we obtain

$$\lim_{(\tilde{x}_0^s, \tilde{x}_0^c) \to (x_0^s, x_0^c)} I_4 = 0.$$

Similarly,

$$\lim_{(\tilde{x}_0^s, \tilde{x}_0^c) \to (x_0^s, x_0^c)} I_2 = 0$$

Thus, from (8.39) it follows

$$\lim_{(\tilde{x}_0^s, \tilde{x}_0^c) \to (x_0^s, x_0^c)} I_1 = 0.$$

To show (8.38), it is enough to show that $DG_2(0,0)$ is an isomorphism from $X_{m_2}^s \times X_{m_2}^c$ onto $X_{m_1}^s \times X_{m_1}^c$. We must show that

$$x^s = \Pi_{m_1}^s (\bar{x}_2^s + \bar{x}_2^c) + \Pi_{m_1}^s Df_2(x_2^s, 0)(\bar{x}_2^s, \bar{x}_2^c)$$
$$x^c = \Pi_{m_1}^c (\bar{x}_2^s + \bar{x}_2^c) + \Pi_{m_1}^c Df_2(x_2^s, 0)(\bar{x}_2^s, \bar{x}_2^c)$$

is uniquely solvable for $(\bar{x}_2^s, \bar{x}_2^c)$. Note that from (8.33) $|m_1 - m_2| \le 5\epsilon$. From (4.5), by choosing ϵ^* sufficiently small such that m_2 is in an η-neighborhood of m_1, $\left(\Pi_{m_1}^s | X_{m_2}^s \right)^{-1}$ and $\left(\Pi_{m_1}^c | X_{m_1}^c \right)^{-1}$ exist. Hence,

$$\bar{x}_2^s = \left(\Pi_{m_1}^s | X_{m_2}^s \right)^{-1} x^s - \left(\Pi_{m_1}^s \Big| X_{m_2}^s \right)^{-1} \left(\Pi_{m_1}^s - \Pi_{m_2}^s \right) \bar{x}_2^c$$

$$- \left(\Pi_{m_1}^s | X_{m_2}^s \right)^{-1} \left(\Pi_{m_1}^s - \Pi_{m_2}^s \right) Df_2(x_2^s, 0)(\bar{x}_2^s, \bar{x}_2^c)$$

$$\bar{x}_2^c = \left(\Pi_{m_1}^c | X_{m_2}^c \right)^{-1} x^c - \left(\Pi_{m_1}^c \Big| x_{m_2}^c \right)^{-1} \left(\Pi_{m_1}^c - \Pi_{m_2}^c \right) \bar{x}_2^s$$

$$- \left(\Pi_{m_1}^c | X_{m_2}^c \right)^{-1} \left(\Pi_{m_1}^c - \Pi_{m_2}^c \right) Df_2(x_2^s, 0)(\bar{x}_2^s, \bar{x}_2^c).$$

We may choose ϵ^* sufficiently small so that the right hand side of the above is a linear contraction on $X_{m_2}^s \times X_{m_2}^c$ uniformly with respect to $(x^s, x^c) \in X_{m_1}^s \times X_{m_1}^c$. This completes the proof. $\square$

From Lemma 7.1 one may also represent the center-stable manifold $\tilde{W}^{cs}(\epsilon)$ locally in terms of functions $\tilde{g}^{cs}(m, \tilde{x}^s)$. It is not hard to see, by using the implicit function theorem, that $\tilde{g}^{cs}$ is C^1.

Summarizing all above propositions gives Theorem 8.1.

9. Smoothness of Center-Unstable Manifolds.

All results obtained in Section 8 also hold for center-unstable manifolds. In particular,

Theorem 9.1. *Let $\lambda_1 \in (\lambda, 1)$, $\rho \in (1, 1/\sqrt{\lambda})$, and $\mu \in (0,1)$ such that $\mu\rho < 1/2$. Then there exist positive constants $\tilde{\epsilon}^*$, $\epsilon^* = \epsilon^*(\tilde{\epsilon})$, $\delta^* = \delta^*(\epsilon) < \epsilon$ and $\sigma = \sigma(\epsilon, \delta)$ such that if $\tilde{\epsilon} < \tilde{\epsilon}^*$, $\epsilon < \epsilon^*$, $\delta < \delta^*$, and $\tilde{T}$ satisfies $\|\tilde{T} - T^t\|_1 < \sigma$, then $\tilde{W}^{cu}(\epsilon)$ obtained in Section 6 is a C^1 manifold.*

Generally, a few modifications are needed to adapt the proofs presented in Section 8 to the case of the center unstable manifold. The most significant differences are the definition of the graph transform and the associated spaces. We shall outline the proofs and leave the details to the interested reader.

The center-unstable manifold we obtained in section 6 is the graph of a Lipschitz section over $X^u(\epsilon)$, which is denoted by $W^{cu}(\epsilon)$ and satisfies

$$\tilde{T}(\tilde{W}^{cu}(\epsilon)) \cap \Theta(X^u(\epsilon) \oplus X^s(\epsilon)) = \tilde{W}^{cu}(\epsilon) \tag{9.1}$$

$$\tilde{T} : \tilde{T}^{-1}(\tilde{W}^{cu}(\epsilon)) \cap \tilde{W}^{cu}(\epsilon) \to \tilde{W}^{cu}(\epsilon) \tag{9.2}$$

is a homeomorphism.

Corresponding to $\Sigma_\theta^{cs,l}$, we define

$$\Sigma_\theta^{cu,l} = \{\gamma : W^{cu}(\epsilon) \to J^\theta(X^u \oplus X^c, X^s; 0, 0), \|\gamma\| < \infty\}$$

where $J^\theta(X^u \oplus X^c, X^s; 0, 0)$ is the jet bundle over M with the jet fiber

$$J^\theta(X_m^u \oplus X_m^c, X_m^s; 0, 0)$$
$$= \{j \in J^b(X_m^u \oplus X_m^c, X_m^s; 0, 0) : j \text{ has a representative } g \text{ satisfying } \mathrm{Lip}(g|_U) \le \theta\}$$

Remember that $\mathrm{Lip}(g|_U)$ is same as in Section 8, U is a neighborhood of 0 in $X_m^u \times X_m^c$, and $\theta = \mu\rho < 1/2$. The norm in $\Sigma_\theta^{cu,l}$ is given by

$$\|\gamma\| = \sup\{\|\gamma(m_1 + x_1^u + x_1^s)\| : m_1 + x_1^u + x_1^s \in W^{cu}(\epsilon)\}.$$

In order to show the smoothness of $W^{cu}(\epsilon)$, we need find a candidate for the tangent bundle of $W^{cu}(\epsilon)$ then to show it indeed is the tangent bundle. The outline of the approach is as follows. The first step is to construct $\tilde{\gamma} \in \Sigma_\theta^{cu,l}$ for each $\gamma \in \Sigma_\theta^{cu,l}$ such that $\tilde{\gamma}$ is the image of γ under $D\tilde{T}$ in certain sense, defining a graph transform

$$\mathcal{F}(\gamma) = \tilde{\gamma}.$$

In the case of the center-stable manifold, we constructed the preimage $\tilde{\gamma}$ of γ instead of the image.

91

The next step is to show that $\mathcal{F}$ has a unique fixed point in $\Sigma_\theta^{cu,l}$. The difficulty here is that $\Sigma_\theta^{cu,l}$ is not complete. To overcome this difficulty, we first show that $\mathcal{F}$ is a contraction in $\Sigma_\theta^{cu,l}$, thus the iterations $\mathcal{F}^k(\gamma)(m_1 + x_1^u + x_1^s)$ define a Cauchy sequence in $J^c(m_1)$. Since $J^c(m_1)$ is complete,

$$\mathcal{F}^k(\gamma)(m_1 + x_1^u + x_1^s) \text{ converges to some } \gamma_0(m_1 + x_1^u + x_1^s) \text{ in } J^c(m_1) \text{ as } k \to 0.$$

On the other hand, the local coordinate representative of $W^{cu}(\epsilon)$, (see Lemma 6.2,)

$$m_0 + x^u + x^s + f(x^u, x^c)$$

produces $\tilde{\gamma}_0 \in \Sigma_\theta^{cu,l}$. It turns out from the invariance of $\tilde{W}^{cu}(\epsilon)$ that $\tilde{\gamma}_0$ is a fixed point of $\mathcal{F}$. Hence, $\tilde{\gamma}_0 = \gamma_0$. The differentiability of f follows from the fact that $\mathcal{F}(J^a) \subset J^a$ and $J^a = J^d$ is closed.

The final step is to show Df is C^0. The idea here is to prove there is a continuous section of a vector bundle, which is a fixed point of $\mathcal{F}$.

Let us first look at how to construct $\mathcal{F}$. For each fixed $x_2 = m_2 + x_2^u + x_2^s \in \tilde{W}^{cu}(\epsilon)$, from (9.1) and (9.2) there is a unique point $x_1 = m_1 + x_1^u + x_1^s \in \tilde{W}^{cu}(\epsilon)$ such that

$$\tilde{T}(x_1) = x_2.$$

For fixed $\gamma \in \Sigma_\theta^{cu,l}$, let

$$j_1 = \gamma(x_1).$$

We choose a Lipschitz representative g_1 from jet j_1 satisfying

$$\|g_1(x^u, x^c) - g_1(\bar{x}^u, \bar{x}^c)\| \le \theta(|x^u - \bar{x}^u| + |x^c - \bar{x}^c|),$$
$$\text{for } (x^u, x^c), (\bar{x}^u, \bar{x}^c) \in B_1^u(0, r_1) \times B_1^c(0, r_1)$$

where $B_1^\alpha(0, r_1), \alpha = u, c$, are closed balls in $X_{m_1}^\alpha$.

We first want to construct a Lipschitz map,

$$\tilde{g}_2 : B_2^u(0, r_2) \times B_2^c(0, r_2) \to B_2^s(0, r_2)$$

with Lip $\tilde{g}_2 \le \theta$ such that for $(\bar{x}^u, \bar{x}^c) \in B_2^u(0, r_2) \times B_2^c(0, r_2)$, there exists a unique $(x^u, x^c) \in B_1^u(0, r_1) \times B_1^c(0, r_1)$ such that

$$D\tilde{T}(x_1)(x^u + x^c + g_1(x^u, x^c)) = \bar{x}^u + \bar{x}^c + \tilde{g}_2(\bar{x}^u, \bar{x}^c). \tag{9.3}$$

To see this, for each $(\bar{x}^u + \bar{x}^c) \in B_2^u(0, r_2) \times B_2^c(0, r_2)$, we define a map E from $B_1^u(0, r_1) \times B_1^c(0, r_1)$ to $X_{m_1}^u \times X_{m_1}^c$ by $E = (E^u, E^c)$, where

$$E^\alpha(x^u, x^c) = \left(\Pi_{m_2}^\alpha D\tilde{T}(x_1)|_{X_{m_1}^\alpha}\right)^{-1} \left(\bar{x}^\alpha - \Pi_{m_2}^\alpha D\tilde{T}(x_1)(x^u + x^c + g_1(x^u, x^c))\right) + x^\alpha$$

for $\alpha = u, c$.

In the same fashion as for E in Lemma 8.3 and 8.4, one may show that there exists $r_1^* > 0$ such that if $r_1 < r_1^*$ then E is well-defined and is a contraction from $B_1^u(0, r_1) \times B_1^c(0, r_1)$ into itself. Thus E has a unique fixed point $(x^u, x^s) = (x^u(\bar{x}^u, \bar{x}^c), x^c(\bar{x}^u, \bar{x}^c))$ which yields a map

$$\bar{x}^s = \tilde{g}_2(\bar{x}^u, \bar{x}^c) = \Pi_{m_2}^s D\tilde{T}(x_1)(x^u + x^c + g_1(x^u + x^s)).$$

As in Lemma 8.4 and Lemma 8.5, one may also show that $\tilde{g}_2$ is Lipschitz with Lip $\tilde{g}_2 \leq \theta$ and $[\tilde{g}_2]$, the jet equivalent class, does not depend on the choice of $g_1 \in j_1$. Hence one has $\tilde{\gamma} \in \Sigma_\theta^{cu,l}$, determined uniquely by γ, where

$$\tilde{\gamma}(x_2) = [\tilde{g}_2].$$

Consequently, one may define a graph transform

$$\mathcal{F}(\gamma) = \tilde{\gamma}.$$

We claim $\mathcal{F}$ is a contraction from $\Sigma_\theta^{cu,l}$ to $\Sigma_\theta^{cu,l}$. There is no significant change in the proof of this claim, comparing with Lemma 8.8.

As we mentioned before, since $\Sigma_\theta^{cu,l}$ is not complete, we cannot directly claim $\mathcal{F}$ has a fixed point in $\Sigma_\theta^{cu,l}$. In order to show that $\mathcal{F}$ has a unique fixed point in $\Sigma_\theta^{cu,l}$ one first observes that the iteration $\mathcal{F}^k(x_2)$ is a Cauchy sequence in $J^c(m_2)$. Since $J^c(m_2)$ is complete, there exists $\gamma_0(x_2) \in J^c(m_2)$ such that

$$\mathcal{F}^k(x_2) \to \gamma_0(x_2)$$

as $k \to \infty$. On the other hand, the local coordinate representative (see Lemma 6.2)

$$m + x^u + x^c + f(x^u, x^c)$$

produces $\tilde{\gamma}_0 \in \Sigma_\theta^{cu,l}$ defined by

$$\tilde{\gamma}_0(x_2) = [f(x_2^u + x^u, x^c) - f(x_2^u, 0)]$$

where f gives the local coordinate representation of $\tilde{W}^{(cu)}(\epsilon)$ at m_2.

From the invariance of $\tilde{W}^{cu}(\epsilon)$, one easily shows that $\tilde{\gamma}_0$ is a fixed point of $\mathcal{F}$, and therefore $\tilde{\gamma}_0 = \gamma_0$. It is not hard to see that if $\gamma(x_2) \in J^a(m_2)$ for all $x_2 \in \tilde{W}^{cu}(\epsilon)$ then

$$\mathcal{F}(\gamma)(x_2) \in J^a(m_2).$$

From the Theorem on Lipschitz jets, $J^a(m_2) = J^d(m_2) \subset J^c(m_2)$ are Banach spaces. Thus, choosing $\gamma \in \Sigma_\theta^{cu,l}$ such that $\gamma(x_2) \in J^a(m_2)$ for all $x_2 \in \tilde{W}^{cu}(\epsilon)$, we obtain

$$\mathcal{F}^k(\gamma) \to \tilde{\gamma}_0$$

and

$$\tilde{\gamma}_0(x_2) \in J^a(m_2) = J^d(m_2),$$

which implies f is differentiable at $(x_2^u, 0)$. The proof of the differentiability of f at (x^u, x^s) follows in the same fashion as in Lemma 8.10.

Finally we want to show that Df is continuous.

As we did in section 8, we shall consider the vector bundle

$$L(X^u \oplus X^c, X^s)$$

set

$$\tilde{\Lambda}_\theta^{cu} = \{\ell : \tilde{W}^{cu}(\epsilon) \to L(X^u \oplus X^c, X^s) \text{ is a } C^0 \text{ section and } ||\ell|| \le \theta\}.$$

Endow $\tilde{\Lambda}_\theta^{cu}$ with norm

$$||\ell|| = \sup\{||\ell(x_1)|| : x_1 \in \tilde{W}^{cu}(\epsilon)\}.$$

It is not hard to check $\tilde{\Lambda}_\theta^{cu}$ is a complete metric space, which may be regarded as a closed subspace of $\Sigma_\theta^{cu,l}$. Here we identify ℓ with $[\ell] \in \Sigma_\theta^{cu,l}$. In the same fashion as in Lemma 8.11 one may show that $\tilde{\mathcal{F}}$ map Λ_θ^{cu} into itself. Observe that $\tilde{\mathcal{F}}$ is a contraction in $\tilde{\Lambda}_\theta^{cu}$ and therefore has a unique fixed point in $\tilde{\Lambda}_\theta^{cu}$, which yields

$$m_2 + x_2^u + f_{m_2}(x_2^u, 0) \to Df(x_2^u, 0)$$

is continuous. Using the same approach as Lemma 8.13 one may show that $Df(x^u, x^c)$ is continuous in (x^u, x^c). Using the Implicit Function Theorem, one may prove that $\tilde{g}^{cu}$ given in Lemma 6.1 is C^1.

10. Persistence of Invariant Manifold.

The kep point here is that the center-unstable and center-stable manifolds constructed earlier are transversal. Actually, the Lipschitz nature of these manifolds provides the intersection, but we wish to also prove its smoothness.

Let $\tilde{M} = \tilde{W}^{cu}(\epsilon) \cap \tilde{W}^{cs}(\epsilon)$. Then we have

Theorem 10.1. *There exists $\epsilon^* > 0$ such that for each $\epsilon < \epsilon^*$ there is $\sigma > 0$ such that if $\|\tilde{T} - T^t\|_1 < \sigma$, then there exists C^1 diffeomorphism*

$$K = K_{\tilde{T}} : M \to \tilde{M}$$

which satisfies

$$\|K_{\tilde{T}} - I\|_{C^1(M,X)} \to 0 \tag{10.1}$$

as $\|\tilde{T} - T^t\|_1 \to 0$. Furthermore $\tilde{M}$ is a C^1 invariant manifold for $\tilde{T}$.

This theorem implies that $\tilde{M}$ is C^1 close to M.

Proof. We first show that $\tilde{M}$ is invariant. Take $\tilde{m} \in \tilde{M}$. By Theorem 7.1, we have that $\tilde{W}^{cs}(\epsilon)$ is forward invariant, hence $\tilde{T}(\tilde{m}) \in \tilde{W}^{cs}(\epsilon)$. On the other hand, from Theorem 6.3, we have

$$\tilde{T}(\tilde{W}^{cu}(\epsilon)) \cap (X^u(\epsilon) \oplus X^s(\epsilon)) = \tilde{W}^{cu}(\epsilon).$$

Hence, by using $\tilde{T}(\tilde{m}) \in \tilde{W}^{cs}(\epsilon)$, which implies that $\tilde{T}(\tilde{m}) \in X^u(\epsilon) \oplus X^s(\epsilon)$, we conclude that

$$\tilde{T}(\tilde{m}) \in \tilde{W}^{cu}(\epsilon).$$

Therefore $\tilde{T}(\tilde{m}) \in \tilde{M}$, which means $\tilde{M}$ is positively invariant.

Next we construct K. Fix $m_1 \in M$, let $\tilde{g}^{cu}$ and $\tilde{g}^{cs}$ be local representatives of $\tilde{h}^{cu}$ and $\tilde{h}^{cs}$, respectively, which are given in Lemma 6.1 and Lemma 7.1. Define a map

$$A : B(m_1, \epsilon\rho) \cap M \times X^u_{m_1}(\rho^{-1}\epsilon) \to X^u_{m_1} \quad \text{by}$$
$$A(m, x^u) = \tilde{g}^{cs}(m, \tilde{g}^{cu}(m, x^u)).$$

We claim that for each $m \in B(m_1, \rho\epsilon) \cap M$, $A(m, x^u)$ has a unique fixed point $x^u = x^u(m) \in X^u(\epsilon\rho^{-1})$, which is C^1 in m.

Recall, from the proofs of Lemma 6.1 and Lemma 7.1, that

$$|\tilde{g}^{cs}(m, 0)| < \delta \tag{10.2}$$

and

$$|\tilde{g}^{cu}(m, 0)| < \delta. \tag{10.3}$$

As in Section 8, we require that $\mu\rho < 1/2$. By (10.2) and (10.3), if we choose $\delta < \frac{\epsilon}{4}$ we have that

$$\max\{|\tilde{g}^{cs}(m,0)|, \; |\tilde{g}^{cu}(m,0)|\} + \mu\epsilon < \rho^{-1}\epsilon.$$

We first show that A is well-defined. Using Lemma 6.1, we obtain

$$|\tilde{g}^{cu}(m,x^u)| \leq |\tilde{g}^{cu}(m,0)| + \rho\mu|x^u|$$
$$\leq |\tilde{g}^{cu}(m,0)| + \mu\epsilon < \rho^{-1}\epsilon$$

which implies that A is well-defined. Furthermore,

$$|A(m,x^u)| < \rho^{-1}\epsilon.$$

Hence for each m, $A(m,\cdot)$ is a map from $X^u_{m_1}(\rho^{-1}\epsilon)$ into itself.

Next we have

$$|A(m,x^u) - A(m,\bar{x}^u)| < (\rho u)^2|x^u - \bar{x}^u|.$$

Thus, $A(m,\cdot)$ is a contraction from $X^u_{m_1}(\rho^{-1}\epsilon)$ into itself. From sections 8 and 9, $\tilde{g}^{cs}$ and $\tilde{g}^{cu}$ are C^1, hence $A(m,x^u)$ is C^1 in (m,x^u). Hence by the uniform contraction mapping theorem, $A(m,\cdot)$ has a unique fixed point $x^u = x^u(m)$ which is C^1 in m. This completes the proof of the claim. Moreover,

$$|x^u(m)| \leq \frac{1}{1-(\rho u)^2}\left(|\tilde{g}^{cs}(m,0)| + \mu\rho|\tilde{g}^{cu}(m,0)|\right).$$

Let $x^s(m) = \tilde{g}^{cu}(m,x^u(m))$ and

$$\tilde{m} = m + \Pi^u_m x^u(m) + \Pi^s_m x^s(m). \tag{10.4}$$

Observe that

$$\{\tilde{m}\} = \left\{m + \tilde{x}^u + \tilde{h}^{cu}(m,\tilde{x}^u) : \tilde{x}^u \in X^u_m(\epsilon)\right\} \cap \left\{m + \tilde{x}^s + \tilde{h}^{cs}(m,\tilde{x}^s) : \tilde{x}^s \in X^s_m(\epsilon)\right\}$$

provided that ϵ^* is sufficiently small. This implies that $\tilde{m}$ is uniquely determined by m and independent of the choice of m_1. Hence, we have a bijection from M onto $\tilde{M}$

$$\tilde{m} = K(m).$$

Since Π^α_m is C^1 in m and $x^u(m)$ and $x^s(m)$ are C^1, K is a C^1 bijection. To show the convergence (10.1), from (10.4) we have

$$|K(m) - m| = |\Pi^u_m x^u(m)| + |\Pi^s_m x^s(m)|$$
$$\leq (1+\eta)(|x^u(m)| + |x^s(m)|)$$
$$\leq 2(1+\eta)\epsilon.$$

Estimating $Dx^u(m)$ and $Dx^s(m)$, we find

$$\|Dx^u(m)\|$$
$$\leq \frac{\rho\mu}{1-\rho\mu}$$

and

$$\|Dx^s(m)\| \leq \rho\mu + \frac{(\rho\mu)^2}{1-\rho\mu}.$$

Hence

$$|DK(m) - I|$$
$$\leq \|D\Pi_m^u\|\,|x^u(m)| + \|D\Pi_m^s\|\,|x^s(m)|$$
$$+ (1+\eta)(\|Dx^u(m)\| + \|Dx^s(m)\|)$$
$$\leq O(\epsilon) + (1+\eta)\left(\rho\mu + (\frac{2(\rho\mu)^2}{1-\rho\mu})\right).$$

We can take ϵ and μ as small as desired by choosing σ sufficiently small. This establishes (10.1) and also implies that $\tilde{M}$ is a C^1 manifold and that K is a diffeomorphism. This completes the proof. $\square$

Proposition 10.2. $\tilde{T} : \tilde{M} \to \tilde{M}$ *is a C^1 diffeomorphism.*

Proof. Theorem 10.1 implies that $\tilde{M}$ is compact. Since T^t is a diffeomorphism from M onto M, again using Proposition 10.1, we have $K \circ T^t \circ K^{-1} : \tilde{M} \to \tilde{M}$ is a diffeomorphism and $\|K \circ T^t \circ K^{-1} - \tilde{T}\|_1 \to 0$ as $\|\tilde{T} - T^t\|_1 \to 0$. Therefore, as long as $\|\tilde{T} - T^t\|_1$ is small enough, $\tilde{T} : \tilde{M} \to \tilde{M}$ is a diffeomorphism . $\square$

From Proposition 6.12 it follows that for each $m_0 + x_0^u + x_0^s \in \tilde{W}^{cu}(\epsilon)$ there is a unique sequence $m_k + x_k^u + x_k^s \in \tilde{W}^{cu}(\epsilon)$ for $k \geq 1$ such that for $k = 1, 2, \cdots$,

$$\tilde{T}(m_k + x_k^u + x_k^s) = m_{k-1} + x_{k-1}^u + x_{k-1}^s.$$

That is, $\tilde{T}^{-k}(m_0 + x_0^u + x_0^s)$ is well-defined for each $m_0 + x_0^u + x_0^s \in \tilde{W}^{cu}(\epsilon)$ in the tubular neighborhood $\Theta(X^u(\epsilon) \oplus X^s(\epsilon))$.

Theorem 10.3.
(i) For each $m + x^u + x^s \in \tilde{W}^{cs}(\epsilon)$,

$$\lim_{k\to\infty} d(\tilde{T}^k(m + x^u + x^s), \tilde{M}) = 0, \text{ uniformly on } \tilde{W}^{cs}(\epsilon) \qquad (10.5)$$

(i) For each $m + x^u + x^s \in \tilde{W}^{cu}(\epsilon)$,

$$\lim_{k\to\infty} d(\tilde{T}^{-k}(m + x^u + x^s), \tilde{M}) = 0, \text{ uniformly on } \tilde{W}^{cu}(\epsilon) \qquad (10.6)$$

where $d(x, \tilde{M}) = \inf_{\tilde{m} \in \tilde{M}} |x - \tilde{m}|$.

Proof. We first consider (10.5). For each $m + x^u + x^s \in \tilde{W}^{cs}(\epsilon)$. Let $m_k + x_k^u + x_k^s = \tilde{T}^k(m + x^u + x^s)$. Observe that $m_k + x_k^u + x_k^s \in \mathcal{A}_k$ where $\mathcal{A}_k$ is given in Proposition 6.10. (6.46) gives

$$|x_k^s - \tilde{h}^{cu}(m_k, x_k^u)| \leq 2\epsilon\lambda_1^k \tag{10.7}$$

We note that

$$K(m_k) = m_k + \bar{x}_k^u + \bar{x}_k^s = m_k + \bar{x}_k^u + \tilde{h}^{cu}(m_k, \bar{x}_k^u) = m_k + \bar{x}_k^s + \tilde{h}^{cs}(m_k, \bar{x}_k^s).$$

Thus, from the fact $m_k + x_k^u + x_k^s \in \tilde{W}^{cs}(\epsilon)$, using Lemma 6.1, we obtain

$$|x_k^u - \bar{x}_k^u| \leq \mu\rho|x_k^s - \bar{x}_k^s| \leq \frac{1}{2}|x_k^s - \bar{x}_k^s|. \tag{10.8}$$

From (10.7), we find

$$|x_k^s - \bar{x}_k^s| - |\tilde{h}^{cu}(m_k, x_k^u) - \tilde{h}^{cu}(m_k, \bar{x}_k^u)| \leq 2\epsilon\lambda_1^k$$

which with (10.8) yields

$$|x_k^s - \bar{x}_k^s| \leq \frac{8}{3}\epsilon\lambda_1^k.$$

Therefore,

$$|K(m_k) - (m_k + x_k^u + x_k^s)| \leq 4\epsilon\lambda_1^k.$$

This completes the proof of (10.5). Similarly, one may show that (10.6) holds. The proof is complete. $\square$

11. Persistence of Normal Hyperbolicity.

In Section 10, we showed that the intersection of the center-stable manifold and center-unstable manifold is a C^1 compact connected invariant manifold. The basic idea to obtain normal hyperbolicity for the perturbed manifold $\tilde{M}$ is to construct stable and unstable bundles from the tangent bundles of the center-stable and center-unstable manifolds by finding projection operators.

Theorem 11.1. *There exists $\sigma > 0$ such that if $\tilde{T}$ satisfies $\|\tilde{T} - T^t\| \leq \sigma$, them $\tilde{M}$ is a normally hyperbolic invariant manifold.*

We first review some basic results from previous sections. We shall again denote the center-stable and center-unstable manifolds for T^t and $\tilde{T}$, respectively, by

$$W^{cs}(\epsilon), \ W^{cu}(\epsilon), \tilde{W}^{cs}(\epsilon) \ \text{and} \ \tilde{W}^{cu}(\epsilon).$$

In sections 8 and 9, we showed that these manifolds are C^1 and consequently their tangent spaces are given, respectively, by the graphs of

$$\ell^{cs} \in C^0(W^{cs}(\epsilon), L(X^s \oplus X^c, X^u)), \tag{11.1}$$

$$\ell^{cu} \in C^0(W^{cu}(\epsilon), L(X^u \oplus X^c, X^s)), \tag{11.2}$$

$$\tilde{\ell}^{cs} \in C^0(\tilde{W}^{cs}(\epsilon), L(X^s \oplus X^c, X^u)), \tag{11.3}$$

$$\tilde{\ell}^{cu} \in C^0(\tilde{W}^{cu}(\epsilon), L(X^u \oplus X^c, X^s)). \tag{11.4}$$

More precisely, the tangent space of $\tilde{W}^{cs}(\epsilon)$, for example, at point $x_1 = m_1 + x_1^u + x_1^s \in \tilde{W}^{cs}(\epsilon)$ is given by

$$T_{x_1}\tilde{W}^{cs}(\epsilon) = \{x^s + x^c + \tilde{\ell}_{x_1}^{cs}(x^s, x^c) : \ x^s + x^c \in X_{m_1}^s \oplus X_{m_1}^c\}.$$

With the norm as before we have that $\|\ell^{cs}\|, \|\ell^{cu}\|, \|\tilde{\ell}^{cs}\|, \|\tilde{\ell}^{cu}\| \leq \theta = \mu\rho < 1/2$.

From Proposition 6.12, $\tilde{T} : \tilde{W}^{cu}(\epsilon) \cap \tilde{T}^{-1}(\tilde{W}^{cu}(\epsilon)) \to \tilde{W}^{cu}(\epsilon)$ is a homeomorphism. Furthermore, since $\tilde{W}^{cu}(\epsilon)$ is a C^1 manifold, one may easily show that $\tilde{T}$ is a C^1 diffeomorphism.

We shall establish Theorem 11.1 by first constructing projection operators. Throughout this section, for each $m \in M$ we set $\tilde{m} = K(m)$, where K is the diffeomorphism obtained in section 10. From (10.4), we notice that $\tilde{m} = m + x^u + x^s$, where $x^u + x^s \in X_m^u(\epsilon) \oplus X_m^u(\epsilon)$.

Our main goal is to find the invariant stable and unstable subspaces. We first try to use the tangent spaces $T_{\tilde{m}}\tilde{W}^{cs}(\epsilon)$ and $T_{\tilde{m}}\tilde{W}^{cu}(\epsilon)$ as a coordinate system to represent each point in $X = X_m^u \oplus X_m^s \oplus X_m^c$. For each $x^u + x^s + x^c \in X_m^u \oplus X_m^s \oplus X_m^c$, we consider the equation

$$x^u + x^s + x^c = \bar{x}^s + \bar{x}^c + \tilde{\ell}_{\tilde{m}}^{cs}(\bar{x}^s, \bar{x}^c) + \bar{x}^u + \tilde{\ell}_{\tilde{m}}^{cu}(\bar{x}^u, 0) \tag{11.5}$$

where $\bar{x}^\alpha \in X_m^\alpha$, $\alpha = u, s, c$. We denote the right hand side of (11.5) by

$$\tilde{\psi}_m^u(\bar{x}^u + \bar{x}^s + \bar{x}^c).$$

Thus we may define a map $\tilde{\psi}^u$ from M to $L(X^u \oplus X^s \oplus X^c)$ by $\tilde{\psi}^u(m) = \tilde{\psi}_m^u$, where $L(X^u \oplus X^s \oplus X^c)$ is the vector bundle over M with fiber $L(X_m^u \oplus X_m^s \oplus X_m^c)$. We should point out that $\bar{x}^u + \tilde{\ell}_m^{cu}(\bar{x}^u, 0)$ is transversal to $T_{\tilde{m}}\tilde{W}^{cs}(\epsilon)$.

The next result shows that (11.5) has a unique solution $\bar{x}^u + \bar{x}^s + \bar{x}^c$.

Lemma 11.2. *$\tilde{\psi}_m^u$ is a bounded linear isomorphism from $X_m^u \oplus X_m^s \oplus X_m^c$ onto itself and $\tilde{\psi}^u \in C^0(M, \ L(X^u \oplus X^s \oplus X^c))$*

Proof. From the fact that $\tilde{\ell}_{\tilde{m}}^{cs}$ and $\tilde{\ell}_{\tilde{m}}^{cu}$ are bounded linear operators and $||\tilde{\ell}_{\tilde{m}}^{c\alpha}|| \leq \rho\mu < 1/2$, $\alpha = u, s$, we obtain that $\tilde{\psi}_m^u$ is a bounded linear operator and satisfies

$$||\tilde{\psi}_m^u|| \leq (1 + \mu\rho) \leq \frac{3}{2}$$

$$|(I - \tilde{\psi}_m^u)(\bar{x}^u + \bar{x}^s + \bar{x}^c)| \leq \mu\rho(|\bar{x}^u| + |\bar{x}^s| + |\bar{x}^c|)$$

$$\leq \frac{1}{2}(|\bar{x}^u| + |\bar{x}^s| + |\bar{x}^c|)$$

which implies that $\tilde{\psi}_m^u$ is a bounded linear isomorphism. From (11.3) and (11.4) it follows that $\tilde{\psi}^u \in C^0(M, L(X^u \oplus X^s \oplus X^c))$. The proof is complete. □

Similarly, we define a map $\tilde{\psi}_m^s$ from $X_m^u \oplus X_m^s \oplus X_m^c$ to $X_m^u \oplus X_m^s \oplus X_m^c$ by

$$\tilde{\psi}_m^s(\bar{x}^u + \bar{x}^s + \bar{x}^c) = \bar{x}^u + \bar{x}^c + \tilde{\ell}_{\tilde{m}}^{cu}(\bar{x}^u, \bar{x}^c) + \bar{x}^s + \tilde{\ell}_{\tilde{m}}^{cs}(\bar{x}^s, 0)$$

and define a map $\tilde{\psi}^s$ from M to $L(X^u \oplus X^s \oplus X^c)$ by $\tilde{\psi}^s(m) = \tilde{\psi}_m^s$.

Then, we may show

Lemma 11.3. *$\tilde{\psi}_m^s$ is a bounded linear isomorphism from $X_m^u \oplus X_m^s \oplus X_m^c$ onto itself and $\tilde{\psi}^s \in C^0(M, L(X^u \oplus X^s \oplus X^c))$.*

Next we define the followings sets of projection operators

$$\Lambda^u = \{\tilde{P}^u \in C^0(M, L(X))| \text{ for } m \in M, \ker(\tilde{P}^u(m)) = T_{\tilde{m}}\tilde{W}^{cs}(\epsilon),$$

$$(\tilde{P}^u(m))^2 = \tilde{P}^u(m) \text{ and } R(\tilde{P}^u(m)) \subset T_{\tilde{m}}\tilde{W}^{cu}(\epsilon)\}$$

and

$$\Lambda^s = \{\tilde{P}^s \in C^0(M, L(X))| \text{ for } m \in M, \ker(\tilde{P}^s(m)) = T_{\tilde{m}}\tilde{W}^{cu}(\epsilon),$$

$$(\tilde{P}^s(m))^2 = \tilde{P}^s(m) \text{ and } R(\tilde{P}^s(m)) \subset T_{\tilde{m}}\tilde{W}^{cs}(\epsilon)\}.$$

It is easy to see that Λ^u and Λ^s are closed subsets of $C^0(M, L(X))$ under the usual norm. Let $L(X, X^u \oplus X^s \oplus X^c)$ be the vector bundle over M with fiber $L(X, X_m^u \oplus X_m^s \oplus X_m^c)$. We have

Lemma 11.4. Λ^u and Λ^s are not empty.

Proof. We first define a section ω of the bundle $L(X, X^u \oplus X^s \oplus X^c)$ by

$$\omega(m)x = \Pi^u_m x + \Pi^s_m x + \Pi^c_m x.$$

The proof of lemma 4.2 implies that ω is C^1. We also define a section $\mathcal{N}^u$ of the bundle $L(X^u \oplus X^s \oplus X^c, X)$ by

$$\mathcal{N}^u(m)(x^u + x^s + x^c) = \bar{x}^u + \tilde{\ell}^{cu}_{\tilde{m}}(\bar{x}^u, 0)$$

where $x^\alpha \in X^\alpha_m$, $\alpha = u, s, c$, and $\bar{x}^u$ is determined by $\bar{x}^u + \bar{x}^s + \bar{x}^c = (\tilde{\psi}^u_m)^{-1}(x^u + x^s + x^c)$. From Lemma 11.2 it follows that $\mathcal{N}^u$ is well-defined and $\mathcal{N}^u \in C^0(M, L(X^u \oplus X^s \oplus X^c, X))$. Observe that $\mathcal{N}^u(m)(x^u + x^s + x^c) \in T_{\tilde{m}}\tilde{W}^{cu}(\epsilon)$. Let $\tilde{P}^u = \mathcal{N}^u\omega$. It is easy to verify $\tilde{P}^u \in \Lambda^u$. Hence $\Lambda^u \neq \phi$.

Similarly, we define a section $\mathcal{N}^s$ of $L(X^u \oplus X^s \oplus X^c, X)$ by

$$\mathcal{N}^s(m)(x^u + x^s + x^c) = \bar{x}^s + \tilde{\ell}^{cs}_{\tilde{m}}(\bar{x}^s, 0)$$

where $\bar{x}^s$ is determined from $\bar{x}^u + \bar{x}^s + \bar{x}^c = (\tilde{\psi}^s_m)^{-1}(x^u + x^s + x^c)$. Then $\mathcal{N}^s\omega \in \Lambda^s$. This completes the proof. $\square$

Next we want to find $\tilde{\Pi}^u \in \Lambda^u$ satisfying the invariance property

$$\tilde{\Pi}^u(m_1)D\tilde{T}(\tilde{m}_0)\tilde{\Pi}^u(m_0) = D\tilde{T}(\tilde{m}_0)\tilde{\Pi}^u(m_0) \qquad (11.6)$$

where $m_1 = K^{-1}(\tilde{m}_1)$ and $\tilde{m}_1 = \tilde{T}(\tilde{m}_0)$, which leads us to define the following transformation.

For each $x^u + x^s + x^c \in X^u_{m_1} \oplus X^s_{m_1} \oplus X^c_{m_1}$, by Lemma 11.2, there exists a unique $\bar{x}^u + \bar{x}^s + \bar{x}^c \in X^u_{m_1} \oplus X^s_{m_1} \oplus X^c_{m_1}$ such that

$$x^u + x^s + x^c = \bar{x}^s + \bar{x}^c + \tilde{\ell}^{cs}_{\tilde{m}_1}(\bar{x}^s, \bar{x}^c) + \bar{x}^u + \tilde{\ell}^{cu}_{\tilde{m}_1}(\bar{x}^u, 0). \qquad (11.7)$$

Note that $x^c = \bar{x}^c$. For $\tilde{P}^u \in \Lambda^u$ and m_1, we define a map $P^u(m_1)$ from X to X by

$$P^u(m_1) = D\tilde{T}(\tilde{m}_0)\tilde{P}^u(m_0)(D\tilde{T}(\tilde{m}_0)|_{T_{\tilde{m}_0}\tilde{W}^{cu}(\epsilon)})^{-1}\mathcal{N}^u(m_1)\omega(m_1),$$

where $m_0 = K^{-1} \circ \tilde{T}^{-1} \circ K(m_1)$. Then $P^u(m_1)$ satisfies

Lemma 11.5.

  (i) $P^u(m_1)$ is well-defined;
  (ii) $\ker(P^u(m_1)) = T_{\tilde{m}_1}\tilde{W}^{cs}(\epsilon)$;
  (iii) $R(P^u(m_1)) \subset T_{\tilde{m}_1}\tilde{W}^{cu}(\epsilon)$;
  (iv) $(P^u(m_1))^2 = P^u(m_1)$.

Proof. As mentioned before, $\tilde{T} : \tilde{W}^{cu}(\epsilon) \cap \tilde{T}^{-1}(\tilde{W}^{cu}(\epsilon)) \rightarrow \tilde{W}^{cu}(\epsilon)$ is a C^1 diffeomorphism, $D\tilde{T}(\tilde{m}_0)$ is an isomorphism from $T_{\tilde{m}_0}\tilde{W}^{cu}(\epsilon)$ to $T_{\tilde{m}_1}\tilde{W}^{cu}(\epsilon)$, where $\tilde{m}_1 = \tilde{T}(\tilde{m}_0)$. Hence $P^u(m_1)$ is well-defined. Clearly, $P^u(m_1)$ is a bounded linear operator from X to X. To show (ii), we first prove that $\ker(P^u(m_1)) \subset T_{\tilde{m}_1}\tilde{W}^{cs}(\epsilon)$. Let $P^u(m_1)(x^u + x^s + x^c) = 0$. Namely,

$$D\tilde{T}(\tilde{m}_0)\tilde{P}^u(m_0)\left(D\tilde{T}(\tilde{m}_0)|_{T_{\tilde{m}_0}\tilde{W}^{cu}(\epsilon)}\right)^{-1}(\bar{x}^u + \tilde{\ell}^{cu}_{\tilde{m}_1}(\bar{x}^u, 0)) = 0.$$

Because $\tilde{P}^u \in \Lambda^u$, $R(\tilde{P}^u(m_0)) \subset T_{\tilde{m}_0}\tilde{W}^{cu}(\epsilon)$. Thus, because $D\tilde{T}(\tilde{m}_0)|_{T_{\tilde{m}_0}\tilde{W}^{cu}(\epsilon)}$ is an isomorphism, we have

$$\tilde{P}^u(m_0)\left(D\tilde{T}(\tilde{m}_0)|_{T_{\tilde{m}_0}\tilde{W}^{cu}(\epsilon)}\right)^{-1}(\bar{x}^u + \tilde{\ell}^{cu}_{\tilde{m}_1}(\bar{x}^u, 0)) = 0.$$

Note that $\ker(\tilde{P}^u(m_0)) = T_{\tilde{m}_0}\tilde{W}^{cs}(\epsilon)$. Hence,

$$\left(D\tilde{T}(\tilde{m}_0)|_{T_{\tilde{m}_0}W^{cu}(\epsilon)}\right)^{-1}(\bar{x}^u + \tilde{\ell}^{cu}_{\tilde{m}_1}(\bar{x}^u, 0)) \in T_{\tilde{m}_0}\tilde{W}^{cs}(\epsilon) \cap T_{\tilde{m}_0}\tilde{W}^{cu}(\epsilon) = T_{\tilde{m}_0}\tilde{M},$$

which yields

$$(\bar{x}^u + \tilde{\ell}^{cu}_{\tilde{m}_1}(\bar{x}^u, 0)) \in T_{\tilde{m}_1}\tilde{M}.$$

Observe that for any $x^u + x^s + x^c \in T_{\tilde{m}_1}\tilde{M}$, since $\|\tilde{\ell}^{cu}_{\tilde{m}_1}\|, \|\tilde{\ell}^{cs}_{\tilde{m}_1}\| < \mu\rho$, we have

$$|x^u| + |x^s| \le \frac{2\rho\mu}{1 - \rho\mu}|x^c|.$$

Therefore, $\bar{x}^u = 0$.

From (11.7)

$$x^u + x^s + x^c = \bar{x}^s + \bar{x}^c + \tilde{\ell}^{cs}_{\tilde{m}_1}(\bar{x}^s, \bar{x}^c) \in T_{\tilde{m}_1}\tilde{W}^{cs}(\epsilon).$$

On the other hand, from the definition of $P^u(m_1)$ it is easy to see that $T_{\tilde{m}_1}\tilde{W}^{cs}(\epsilon) \subset \ker(P^u(m_1))$. Therefore (ii) holds. It is quite straightforward to check (iii). To prove (iv), let $y = P^u(m_1)x$ and $z = (I - \mathcal{N}^u(m_1)\omega(m_1))y$ so that $z \in R(I - \mathcal{N}^u(m_1)\omega(m_1)) \subset T_{\tilde{m}_1}\tilde{W}^{cu}(\epsilon)$. But by (iii), $y \in T_{\tilde{m}_1}\tilde{W}^{cu}(\epsilon)$ and $\mathcal{N}^u(m_1)\omega(m_1))y \in T_{\tilde{m}_1}\tilde{W}^{cu}(\epsilon)$. So $z \in T_{\tilde{m}_1}\tilde{M}$, which implies

$$\tilde{P}^u(m_0)(D\tilde{T}(\tilde{m}_0)|_{T_{\tilde{m}_0}\tilde{W}^{cu}(\epsilon)})^{-1}z = 0.$$

Therefore,

$$
\begin{aligned}
&(P^u(m_1))^2 x \\
&= D\tilde{T}(\tilde{m}_0)\tilde{P}^u(m_0)(D\tilde{T}(\tilde{m}_0)|_{T_{\tilde{m}_0}\tilde{W}^{cu}(\epsilon)})^{-1}(y-z) \\
&= D\tilde{T}(\tilde{m}_0)\tilde{P}^u(m_0)(D\tilde{T}(\tilde{m}_0)|_{T_{\tilde{m}_0}\tilde{W}^{cu}(\epsilon)})^{-1}D\tilde{T}(\tilde{m}_0)\tilde{P}^u(m_0)(D\tilde{T}(\tilde{m}_0)|_{T_{\tilde{m}_0}\tilde{W}^{cu}(\epsilon)})^{-1} \\
&\qquad \mathcal{N}^u(m_1)\omega(m_1)x \\
&= P^u(m_1)x
\end{aligned}
$$

The proof is complete. □

Note that since $\tilde{T}$ is a diffeomorphism on $\tilde{M}, m_1$ can be taken to be any point of M. Thus, P^u is a map from M to $L(X)$. From continuity of $D\tilde{T}$, $\tilde{P}^u$, $\mathcal{N}^u$, and ω it follows that $P^u \in C^0(M, L(X))$. Summarizing above discussion gives next result.

Proposition 11.6. $P^u \in \Lambda^u$.

Thus, we may define a map $\mathcal{F}^u$ from Λ^u to Λ^u by

$$
\mathcal{F}^u(\tilde{P}^u) = P^u.
$$

In order to show $\mathcal{F}^u$ has a unique fixed point in Λ^u, we shall introduce an equivalent metric in Λ^u under which $\mathcal{F}^u$ is a contraction.

Let $m \in M$. For $x^u + x^s + x^c \in X_m^u \oplus X_m^s \oplus X_m^c$, by Lemma 11.2, there exists a unique $\bar{x}^u + \bar{x}^s + \bar{x}^c \in X_m^u \oplus X_m^s \oplus X_m^c$ such that

$$
x^u + x^s + x^c = \bar{x}^s + \bar{x}^c + \tilde{\ell}_{\tilde{m}}^{cs}(\bar{x}^s, \bar{x}^c) + \bar{x}^u + \tilde{\ell}_{\tilde{m}}^{cu}(\bar{x}^u, 0).
$$

Let $\tilde{P}_1^u, \tilde{P}_2^u \in \Lambda^u$, since $R(\tilde{P}_i^u(m)) \subset T_{\tilde{m}}\tilde{W}^{cu}(\epsilon), i = 1, 2$, there exists a unique $\hat{x}^u + \hat{x}^c \in X_m^u \oplus X_m^c$ such that

$$
\begin{aligned}
&(\tilde{P}_1^u(m) - \tilde{P}_2^u(m))(x^u + x^s + x^c) \\
&= (\tilde{P}_1^u(m) - \tilde{P}_2^u(m))(\bar{x}^u + \tilde{\ell}_{\tilde{m}}^{cu}(\bar{x}^u, 0)) = \hat{x}^u + \hat{x}^c + \tilde{\ell}_{\tilde{m}}^{cu}(\hat{x}^u, \hat{x}^c).
\end{aligned}
\tag{11.8}
$$

Define

$$
\delta_u(\tilde{P}_1^u(m), \tilde{P}_2^u(m)) \equiv \sup_{|\bar{x}^u|=1} (|\hat{x}^u| + |\hat{x}^c|).
\tag{11.9}
$$

Then we have

Lemma 11.7. *There exists a constant $C > 0$ such that*

$$
\frac{1}{C}\|\tilde{P}_1^u(m) - \tilde{P}_2^u(m)\| \le \delta_u(\tilde{P}_1^u(m), \tilde{P}_2^u(m)) \le C\|\tilde{P}_1^u(m) - \tilde{P}_2^u(m)\|, \text{ for } m \in M.
$$

Proof. From(11.8) and the fact that $||\tilde{\ell}_{\tilde{m}}^{cu}|| \leq \mu\rho < 1/2$, we obtain

$$(1 - \mu\rho)(|\hat{x}^u| + |\hat{x}^c|) \leq ||\tilde{P}_1^u(m) - \tilde{P}_2^u(m)||(1 + \mu\rho)|\bar{x}^u|$$

which yields

$$\delta_u(\tilde{P}_1^u(m), \tilde{P}_2^u(m)) \leq 3||\tilde{P}_1^u(m) - \tilde{P}_2^u(m)||.$$

On the other hand, for any $x \in X$, writing it as (11.7) and using the fact that $\tilde{P}_1^u, \tilde{P}_2^u \in \Lambda^u$, there exist $\hat{x}^u + \hat{x}^c \in X_m^u \oplus X_m^c$ such that

$$(\tilde{P}_1^u(m) - \tilde{P}_2^u(m))x = (\tilde{P}_1^u(m) - \tilde{P}_2^u(m))(\bar{x}^u + \ell_{\tilde{m}}^{cu}(\bar{x}^u, 0))$$
$$= \hat{x}^u + \hat{x}^c + \hat{\ell}_{\tilde{m}}^{cu}(\hat{x}^u, \hat{x}^c).$$

Thus, by definition (11.9),

$$|(\tilde{P}_1^u(m) - \tilde{P}_2^u(m))x| \leq (1 + \mu\rho)(|\hat{x}^u| + |\hat{x}^c|)$$
$$\leq \frac{3}{2}|\bar{x}^u|\delta_u(\tilde{P}_1^u(m), \tilde{P}_2^u(m)).$$

Observe that $\bar{x}^u = (\Pi_m^u \mathcal{N}^u \omega)x$. Hence

$$|\bar{x}^u| \leq ||\Pi_m^u|| ||\mathcal{N}^u(m)|| ||\omega(m)|| |x| \leq C|x|$$

for some positive constant C independent of $m \in M$. Choosing a possiblely larger constant C completes the proof. $\square$

Set

$$d_u(\tilde{P}_1^u, \tilde{P}_2^u) \equiv \sup_{m \in M} \delta_u(\tilde{P}_1^u(m), \tilde{P}_2^u(m)). \tag{11.10}$$

Then Lemma 11.7 implies that (11.10) defines an equivalent metric in Λ^u.

Proposition 11.8. *There exists $\sigma > 0$ such that if $||\tilde{T} - T^t||_1 \leq \sigma$, then $\mathcal{F}^u : \Lambda^u \to \Lambda^u$ is a contraction under the metric (11.10).*

Proof. For $\bar{x}^u \in X_{\tilde{m}_1}^u$, let

$$x^u + x^s + x^c = \left(D\tilde{T}(\tilde{m}_0)|_{T_{\tilde{m}_0}\tilde{W}^{cu}(\epsilon)}\right)^{-1} \left(\bar{x}^u + \tilde{\ell}_{\tilde{m}_1}^{cu}(\bar{x}^u, 0)\right). \tag{11.11}$$

Here $\tilde{m}_1 = \tilde{T}(\tilde{m}_0)$ and $x^\alpha \in X_{\tilde{m}_0}^\alpha$ for $\alpha = u, s, c$. Thus,

$$\bar{x}^u + \tilde{\ell}_{\tilde{m}_1}^{cu}(\bar{x}^u, 0) = D\tilde{T}(\tilde{m}_0)(x^u + x^s + x^c). \tag{11.12}$$

We also observe, from the invariance, that

$$x^s = \tilde{\ell}_{\tilde{m}_0}^{cu}(x^u, x^c). \tag{11.13}$$

As in (11.7), by Lemma 11.2, there exists $\hat{x}^u + \hat{x}^s + \hat{x}^c \in X_{m_0}^u \oplus X_{m_0}^s \oplus X_{m_0}^c$ such that

$$x^u + x^s + x^c = \hat{x}^s + \hat{x}^c + \tilde{\ell}_{\tilde{m}_0}^{cs}(\hat{x}^s, \hat{x}^c) + \hat{x}^u + \tilde{\ell}_{\tilde{m}_0}^{cu}(\hat{x}^u, 0) \qquad (11.14)$$

Note that $x^c = \hat{x}^c$. We claim

$$|\bar{x}^u| \geq \Big(\inf\{|DT^t(m_0)x^u| : |x^u| = 1, x^u \in X_{m_0}^u\} - O(\epsilon) - C\sigma\Big)|\hat{x}^u| \qquad (11.15)$$

To show this claim, we first observe that $\hat{x}^s + x^c + \tilde{\ell}_{\tilde{m}_0}^{cs}(\hat{x}^s, x^c) \in T_{\tilde{m}_0}\tilde{M}$. In fact, (11.13) implies that $\hat{x}^s + x^c + \tilde{\ell}_{\tilde{m}_0}^{cs}(\hat{x}^s, x^c) + \hat{x}^u + \tilde{\ell}_{\tilde{m}_0}^{cu}(\hat{x}^u, 0) \in T_{\tilde{m}_0}\tilde{W}^{cu}(\epsilon)$, thus

$$\hat{x}^s + x^c + \tilde{\ell}_{\tilde{m}_0}^{cs}(\hat{x}^s, x^c) \in T_{\tilde{m}_0}\tilde{W}^{cu}(\epsilon).$$

Hence, $\hat{x}^s + x^c + \tilde{\ell}_{\tilde{m}_0}^{cs}(\hat{x}^s, x^c) \in T_{\tilde{m}_0}\tilde{M}$.

A simple calculation gives

$$|\hat{x}^s| + |\tilde{\ell}_{\tilde{m}_0}^{cs}(\hat{x}^s, x^c)| \leq \frac{2\mu\rho}{1 - \mu\rho}|x^c| \leq 2|x^c|. \qquad (11.16)$$

From (10.4) we have for $i = 0, 1$,

$$|m_i - \tilde{m}_i| \leq 2\epsilon. \qquad (11.17)$$

Thus, One may obtain, for some positive constant C,

$$|m_1 - T^t(m_0)| \leq C(\epsilon + \sigma). \qquad (11.18)$$

Applying the projection $\Pi_{m_1}^c$ to (11.12),

$$0 = \Pi_{m_1}^c D\tilde{T}(\tilde{m}_0)(x^u + x^s + x^c).$$

Using (11.14), (11.16), (11.17), (11.18), and an estimate similar to (8.8), we have

$$\begin{aligned}
|DT^t(m_0)x^c| &= |DT^t(m_0)x^c - \Pi_{m_1}^c D\tilde{T}(\tilde{m}_0)(x^u + x^s + x^c)| \\
&\leq |DT^t(m_0)x^c - \Pi_{m_1}^c D\tilde{T}(\tilde{m}_0)x^c| + |\Pi_{m_1}^c D\tilde{T}(\tilde{m}_0)(x^u + x^s)| \\
&\leq (O(\epsilon) + C\sigma)|x^c| + (O(\epsilon) + C\sigma)|\hat{x}^u|
\end{aligned}$$

which yields, by using (H3),

$$|x^c| \leq (O(\epsilon) + C\sigma)|\hat{x}^u| \qquad (11.19)$$

Similarly, applying the projection $\Pi_{m_1}^u$ to (11.12), we obtain

$$|\bar{x}^u| \geq \inf\Big\{|DT^t(m_0)x^u| : |x^u| = 1, x^u \in X_{m_0}^u\Big\}|\hat{x}^u| - (O(\epsilon) + C\sigma)(|x^c| + |\hat{x}^u|)$$

which, with (11.19), gives (11.15). This completes the proof of the claim.

For $\tilde{P}_1^u, \tilde{P}_2^u \in \Lambda^u$ and $m_0 \in M$, let

$$P_i^u = \mathcal{F}^u(\tilde{P}_i^u)(m_1)$$

where $m_1 = K^{-1} \circ \tilde{T} \circ K(m_0)$.

Observe that for $i = 1, 2$,

$$\tilde{P}_i^u(m_0)(I - \tilde{P}_i^u(m_0)) = 0, \text{ and } P_i^u(m_1)(I - P_i^u(m_1)) = 0,$$

which implies

$$R(I - \tilde{P}_i^u(m_0)) \subset T_{\tilde{m}_0} \tilde{W}^{cs}(\epsilon),$$

$$R(I - P_i^u(m_1)) \subset T_{\tilde{m}_1} \tilde{W}^{cs}(\epsilon).$$

Thus,

$$R(\tilde{P}_1^u(m_0) - \tilde{P}_2^u(m_0)) \subset T_{\tilde{m}_0} \tilde{M}, \tag{11.20}$$

and

$$R(P_1^u(m_1) - P_2^u(m_1)) \subset T_{\tilde{m}_1} \tilde{M}. \tag{11.21}$$

For $\bar{x}^u \in X_{m_1}^u$, let

$$x_1^u + x_1^s + x_1^c = (P_1^u(m_1) - P_2^u(m_1))(\bar{x}^u + \tilde{\ell}_{\tilde{m}_1}^{cu}(\bar{x}^u, 0))).$$

From the definition of $\mathcal{F}^u$, using (11.11) and (11.14), we find

$$
\begin{aligned}
x_1^u &+ x_1^s + x_1^c \\
&= D\tilde{T}(m_0)(\tilde{P}_1^u(m_0) - \tilde{P}_2^u(m_0)) \left(D\tilde{T}(\tilde{m}_0)|_{T_{\tilde{m}_0} \tilde{W}^{cu}(\epsilon)} \right)^{-1} (\bar{x}^u + \tilde{\ell}_{\tilde{m}_1}^{cu}(\bar{x}^u, 0)) \\
&= D\tilde{T}(m_0)(\tilde{P}_1^u(m_0) - \tilde{P}_2^u(m_0))(\hat{x}^u + \tilde{\ell}_{\tilde{m}_0}^{cu}(\hat{x}^u, 0)). \tag{11.22}
\end{aligned}
$$

Set

$$\tilde{x}^u + \tilde{x}^s + \tilde{x}^c = (\tilde{P}_1^u(m_0) - \tilde{P}_2^u(m_0))(\hat{x}^u + \tilde{\ell}_{\tilde{m}_0}^{cu}(\hat{x}^u, 0)). \tag{11.23}$$

From (11.20), we have $\tilde{x}^u + \tilde{x}^s + \tilde{x}^c \in T_{\tilde{m}_0} \tilde{M}$, hence,

$$|\tilde{x}^u| + |\tilde{x}^s| \leq \frac{2\rho\mu}{1 - \rho\mu} |\tilde{x}^c| \tag{11.24}$$

and

$$|\tilde{x}^u| + |\tilde{x}^c| \leq \frac{1}{1 - \rho\mu} |\tilde{x}^c|. \tag{11.25}$$

Applying the projection $\Pi^c_{m_1}$ to (11.22), using (11.17), (11.18), (11.24), and an estimate similar to (8.8), we obtain

$$
\begin{aligned}
|x^c_1| &= |\Pi^c_{m_1} D\tilde{T}(\tilde{m}_0)(\tilde{x}^u + \tilde{x}^s + \tilde{x}^c)| \\
&\le \|DT^t(m_0)|_{X^c_{m_0}}\| \, |\tilde{x}^c| + |(DT^t(m_0) - \Pi^c_{m_1} D\tilde{T}(\tilde{m}_0))\tilde{x}^c| + |\Pi^c_{m_1} D\tilde{T}(\tilde{m}_0)(\tilde{x}^u + \tilde{x}^s)| \\
&\le \Big(\|DT^t(m_0)|_{X^c_{m_0}}\| + O(\epsilon) + C\sigma \Big)|\tilde{x}^c| \\
&\le \Big(\|DT^t(m_0)|_{X^c_{m_0}}\| + O(\epsilon) + C\sigma \Big) d_u(\tilde{P}^u_1, \tilde{P}^u_2)|\hat{x}^u|,
\end{aligned}
$$

where (11.9) and (11.23) are used in the last estimate.

Thus, by (11.15), we have

$$
|x^c_1| \le \frac{\Big(\|DT^t(m_0)|_{X^c_{m_0}}\| + O(\epsilon) + C\sigma \Big) d_u(\tilde{P}^u_1, \tilde{P}^u_2)}{\Big(\inf\{|DT^t(m_0)x^u| : |x^u| = 1, x^u \in X^u_{m_0}\} - O(\epsilon) - C\sigma \Big)} |\tilde{x}^u|.
$$

Note that $x^u_1 + x^s_1 + x^c_1 \in T_{\tilde{m}_1}\tilde{M}$. Thus, as in (11.25), we have

$$
|x^u_1| + |x^c_1| \le \frac{1}{1 - \rho\mu}|x^c_1|,
$$

hence,

$$
\frac{|x^u_1| + |x^c_1|}{|\tilde{x}^u|} \le \frac{1}{1 - \rho\mu} \frac{\Big(\|DT^t(m_0)|_{X^c_{m_0}}\| + O(\epsilon) + C\sigma \Big) d_u(\tilde{P}^u_1, \tilde{P}^u_2)}{\Big(\inf\{|DT^t(m_0)x^u| : |x^u| = 1, x^u \in X^u_{m_0}\} - O(\epsilon) - C\sigma \Big)}.
$$

Observe that (H3) implies that

$$
\frac{\|DT^t(m_0)|_{X^c_{m_0}}\|}{\inf\{|DT^t(m_0)x^u| : |x^u| = 1, x^u \in X^u_{m_0}\}} < \lambda.
$$

By choosing μ, ϵ^* and σ sufficiently small, we have for $\tilde{T}$ satisfying $\|\tilde{T} - T^t\|_1 \le \sigma$

$$
\delta_u(P^u_1(m_1), P^u_2(m_1)) \le \lambda_1 d_u(\tilde{P}^u_1, \tilde{P}^u_2).
$$

Hence,

$$
d_u(\mathcal{F}^u(\tilde{P}^u_1), \mathcal{F}^u(\tilde{P}^u_2)) \le \lambda_1 d_u(\tilde{P}^u_1, \tilde{P}^u_2) \tag{11.26}
$$

and the proof is complete. $\square$

Note that Λ^u is complete under the metric (11.10) from Lemma 11.7. Since $\mathcal{F}^u$ is a contraction, by the contraction mapping theorem, we obtain

Proposition 11.9. *There exists $\sigma > 0$ such that if $\|\tilde{T} - T^t\|_1 \leq \sigma$, there exists $\tilde{\Pi}^u \in \Lambda^u$ such that*

$$\tilde{\Pi}^u(K^{-1}(\tilde{m}_1))D\tilde{T}(\tilde{m}_0)\tilde{\Pi}^u(K^{-1}(\tilde{m}_0)) = D\tilde{T}(\tilde{m}_0)\tilde{\Pi}^u(K^{-1}(\tilde{m}_0)).$$

Proposition 11.10. $\|\Pi^u - \tilde{\Pi}^u\|_0 \to 0$ *as* $\|\tilde{T} - T^t\|_1 \to 0$.

Proof. Note that $\mathcal{N}^u\omega \in \Lambda^u$ from the proof of Lemma 11.4. We first estimate $d_{(}\mathcal{N}^u\omega, \mathcal{F}^u(\mathcal{N}^u\omega))$. For each $\tilde{x}^u + \tilde{x}^s + \tilde{x}^c \in X^u_{m_1} \oplus X^s_{m_1} \oplus X^c_{m_1}$, by Lemma 11.2, there exist a unique $\bar{x}^u + \bar{x}^s + \bar{x}^c \in X^u_{m_1} \oplus X^s_{m_1} \oplus X^c_{m_1}$ such that

$$\tilde{x}^u + \tilde{x}^s + \tilde{x}^c = \bar{x}^s + \bar{x}^c + \tilde{\ell}^{cs}_{\tilde{m}_1}(\bar{x}^s, \bar{x}^c) + \bar{x}^u + \tilde{\ell}^{cu}_{\tilde{m}_1}(\bar{x}^u, 0).$$

Note $\tilde{x}^c = \bar{x}^c$. As in (11.11), let

$$x^u + x^s + x^c = \left(D\tilde{T}(\tilde{m}_0)\big|_{T_{\tilde{m}_0}\tilde{W}^{cu}(\epsilon)}\right)^{-1}(\bar{x}^u + \tilde{\ell}^{cu}_{\tilde{m}_1}(\bar{x}^u, 0)),$$

and write $x^u + x^s + x^c$ as in (11.14). Combining (11.16) with (11.15) and (11.19) gives

$$|\hat{x}^s| + |\hat{x}^c| + |\tilde{\ell}^{cs}_{\tilde{m}_0}(\hat{x}^s, \hat{x}^c)| \leq (O(\epsilon) + C\sigma)|\bar{x}^u|. \tag{11.27}$$

Thus,

$$\begin{aligned}
&\left|\left(\mathcal{N}^u(m_1)\omega(m_1) - (\mathcal{F}^u(\mathcal{N}^u\omega))(m_1)\right)(\tilde{x}^u + \tilde{x}^s + \tilde{x}^c)\right| \\
&= \left|\left(\mathcal{N}^u(m_1)\omega(m_1) - (\mathcal{F}^u(\mathcal{N}^u\omega))(m_1)\right)\left(\bar{x}^u + \tilde{\ell}^{cu}_{\tilde{m}_1}(\bar{x}^u, 0)\right)\right| \\
&= |D\tilde{T}(\tilde{m}_0)(\hat{x}^s + \hat{x}^c + \tilde{\ell}^{cs}_{\tilde{m}_0}(\hat{x}^s, \hat{x}^c))| \\
&\leq (O(\epsilon) + C\sigma)|\bar{x}^u|,
\end{aligned}$$

which implies

$$d_u(\mathcal{N}^u\omega, \mathcal{F}^u(\mathcal{N}^u\omega))) \leq O(\epsilon) + C\sigma.$$

Since $\tilde{\Pi}^u$ is the fixed point of the contraction mapping $\mathcal{F}^u$, using (11.26) gives

$$d_u(\mathcal{N}^u\omega, \tilde{\Pi}^u) \leq (\sum_{k=0}^{\infty} \lambda_1^k)(O(\epsilon) + C\sigma) = O(\epsilon) + C\sigma \tag{11.28}$$

with a different constant C.

Next, we estimate $\|\mathcal{N}^u\omega - \Pi^u\|$. From (11.5), we obtain

$$\|(\tilde{\psi}^u(m_1))^{-1} - I\| \leq \frac{\rho\mu}{1 - \rho\mu} \tag{11.29}$$

and $\bar{x}^u + \bar{x}^s + \bar{x}^c = (\tilde{\psi}^u(m_1))^{-1}(\tilde{x}^u + \tilde{x}^s + \tilde{x}^c)$, which implies

$$|\bar{x}^u - \tilde{x}^u| \le \frac{\mu\rho}{1 - \mu\rho}(|\tilde{x}^u| + |\tilde{x}^s| + |\tilde{x}^c|)$$

and

$$|\bar{x}^u| \le \frac{1}{1 - \mu\rho}(|\tilde{x}^u| + |\tilde{x}^s| + |\tilde{x}^c|).$$

Thus,

$$\begin{aligned}
|(\mathcal{N}^u(m_1)\omega(m_1) &- \Pi^u_{m_1})(\tilde{x}^u + \tilde{x}^s + \tilde{x}^c)| \\
&\le |\bar{x}^u + \ell^{cu}_{\tilde{m}_1}(\bar{x}^u, 0) - \tilde{x}^u| \\
&\le |\bar{x}^u - \tilde{x}^u| + \mu\rho|\bar{x}^u| \\
&\le \frac{2\mu\rho}{1 - \mu\rho}(|\tilde{x}^u| + |\tilde{x}^s| + |\tilde{x}^c|) \\
&\le C\frac{2\mu\rho}{1 - \rho\mu}|\tilde{x}^u + \tilde{x}^s + \tilde{x}^c|,
\end{aligned}$$

which implies

$$\|\mathcal{N}^u\omega - \Pi^u\| \le C\frac{2\rho\mu}{1 - \rho\mu}.$$

Hence, using (11.28) and Lemma 11.7, we find

$$\begin{aligned}
\|\tilde{\Pi}^u - \Pi^u\| &\le \|\tilde{\Pi}^u - \mathcal{N}^u\omega\| + \|\mathcal{N}^u\omega - \Pi^u\| \\
&\le C d_u(\tilde{\Pi}^u, \mathcal{N}^u\omega) + \|\mathcal{N}^u\omega - \Pi^u\| \\
&\le O(\epsilon) + C\sigma + \frac{2\mu\rho C}{1 - \mu\rho}.
\end{aligned}$$

For any $\mathcal{E} > 0$, we first choose μ small enough that $\frac{2\mu\rho C}{1-\mu\rho} < \frac{1}{2}\mathcal{E}$, then we may choose ϵ^* and σ sufficiently small such that that if $\|\tilde{T} - T^t\|_1 < \sigma$, then

$$\|\tilde{\Pi}^u - \Pi^u\| \le \mathcal{E}$$

which completes the proof. □

Next we shall show that there is a unique projection map $\tilde{\Pi}^s \in \Lambda^s$ satisfying the invariance property

$$\tilde{\Pi}^s(m_1)D\tilde{T}(\tilde{m}_0) = D\tilde{T}(\tilde{m}_0)\tilde{\Pi}^s(m_0).$$

We cannot find $\tilde{\Pi}^s$ by directly applying the contraction mapping approach to the above equation since $D\tilde{T}(\tilde{m}_0)|_{T_{\tilde{m}_0}\tilde{W}^{cs}(\epsilon)}$ may not be invertible. To overcome this difficulty, we consider the following equivalent equation instead

$$D\tilde{T}(\tilde{m}_0)(\tilde{\Pi}^s(m_0) - I) = (\tilde{\Pi}^s(m_1) - I)D\tilde{T}(\tilde{m}_0).$$

Note that for $\tilde{\Pi}^s \in \Lambda^s$

$$R(\tilde{\Pi}^s(m_0) - I) \subset \ker(\tilde{\Pi}^s(m_0)) = T_{\tilde{m}_0}\tilde{W}^{cu}(\epsilon) \tag{11.30}$$

and

$$R(\tilde{\Pi}^s(m_1) - I) \subset \ker(\tilde{\Pi}^s(m_1)) = T_{\tilde{m}_1}\tilde{W}^{cu}(\epsilon). \tag{11.31}$$

Thus,

$$\tilde{\Pi}^s(m_0) = I + \left(D\tilde{T}(\tilde{m}_0)\big|_{T_{\tilde{m}_0}\tilde{W}^{cu}(\epsilon)}\right)^{-1}(\tilde{\Pi}^s(m_1) - I)D\tilde{T}(\tilde{m}_0)$$

which leads us to define the following transformation: For $\tilde{P}^s \in \Lambda^s$, let

$$P^s(m_0) = I + \left(D\tilde{T}(\tilde{m}_0)\big|_{T_{\tilde{m}_0}\tilde{W}^{cu}(\epsilon)}\right)^{-1}(\tilde{P}^s(m_1) - I)D\tilde{T}(\tilde{m}_0).$$

Lemma 11.11.

    (i) $P^s(m_0)$ *is well-defined;*
    (ii) $\ker(P^s(m_0)) = T_{\tilde{m}_0}\tilde{W}^{cu}(\epsilon);$
    (iii) $R(P^s(m_0)) \subset T_{\tilde{m}_0}\tilde{W}^{cs}(\epsilon);$
    (iv) $(P^s(m_0))^2 = P^s(m_0).$

Proof. . Since $D\tilde{T}(\tilde{m}_0) : T_{\tilde{m}_0}\tilde{W}^{cu}(\epsilon) \to T_{\tilde{m}_1}\tilde{W}^{cu}(\epsilon)$ is an isomorphism, $P^s(m_0)$ is well-defined. Clearly $P^s(m_0)$ is a bounded linear operator. The proofs of (ii) and (iv) are quite straightforward. Let us look at (iii). For each $x^u + x^s + x^c \in X^u_{m_0} \oplus X^s_{m_0} \oplus X^c_{m_0}$, by Lemma 11.3, there exits a unique $\bar{x}^u + \bar{x}^s + \bar{x}^c \in X^u_{m_0} \oplus X^s_{m_0} \oplus X^c_{m_0}$ such that

$$x^u + x^s + x^c = \bar{x}^u + \bar{x}^c + \tilde{\ell}^{cu}_{\tilde{m}_0}(\bar{x}^u, \bar{x}^c) + \bar{x}^s + \tilde{\ell}^{cs}_{\tilde{m}_0}(\bar{x}^s, 0).$$

Note that $x^c = \bar{x}^c$. By (11.31) and (ii) in this lemma, we have

$$P^s(m_0)(x^u + x^s + x^c)$$

$$= P^s(m_0)(\bar{x}^s + \tilde{\ell}^{cs}_{\tilde{m}_0}(\bar{x}^s, 0))$$

$$= \bar{x}^s + \tilde{\ell}^{cs}_{\tilde{m}_0}(\bar{x}^s, 0) - \left(D\tilde{T}(\tilde{m}_0)\big|_{T_{\tilde{m}_0}\tilde{W}^{cu}(\epsilon)}\right)^{-1}(\tilde{P}^s(m_1) - I)D\tilde{T}(\tilde{m}_0)(\bar{x}^s + \tilde{\ell}^{cs}_{\tilde{m}_0}(\bar{x}^s, 0)).$$

Observe that $(\tilde{P}^s(m_1) - I)D\tilde{T}(\tilde{m}_0)(\bar{x}^s + \tilde{\ell}^{cs}_{\tilde{m}_0}(\bar{x}^s, 0)) \in T_{\tilde{m}_1}\tilde{M}$. Hence,

$$P^s(m_0)(\bar{x}^s + \tilde{\ell}^{cs}_{\tilde{m}_0}(\bar{x}^s, 0)) - (\bar{x}^s + \tilde{\ell}^{cs}_{\tilde{m}_0}(\bar{x}^s, 0)) \in T_{\tilde{m}_0}\tilde{M}$$

which implies

$$P^s(m_0)(\bar{x}^s + \tilde{\ell}^{cs}_{\tilde{m}_0}(\bar{x}^s, 0)) \in T_{\tilde{m}_0}\tilde{W}^{cs}(\epsilon).$$

This completes the proof. $\square$

Lemma 11.11 defines a map P^s from M to $L(X)$. The continuity of $D\tilde{T}$ and $\tilde{P}^s$ implies that $P^s \in C^0(M, L(X))$. So we have proved

Lemma 11.12. $P^s \in \Lambda^s$

Define $\mathcal{F}^s : \Lambda^s \to \Lambda^s$ by $\mathcal{F}^s(\tilde{P}^s) = P^s$. As in Proposition 11.8, we shall show that $\mathcal{F}^s$ is a contraction under an equivalent metric. Let $m \in M$. For $x^u + x^s + x^c \in X_m^u \oplus X_m^s \oplus X_m^c$, by Lemma 11.3, there exists a unique $\bar{x}^u + \bar{x}^s + \bar{x}^c \in X_{\bar{m}}^u \oplus X_{\bar{m}}^s \oplus X_{\bar{m}}^c$ such that

$$x^u + x^s + x^c = \bar{x}^u + \bar{x}^c + \tilde{\ell}_{\bar{m}}^{cu}(\bar{x}^u, \bar{x}^c) + \bar{x}^s + \tilde{\ell}_{\bar{m}}^{cs}(\bar{x}^s, 0).$$

Note $x^c = \bar{x}^c$. For $\tilde{P}_1^s$ and $\tilde{P}_2^s \in \Lambda^s$, since $R(\tilde{P}_i^s(m)) \subset T_{\bar{m}}\tilde{W}^{cs}(\epsilon)$, there exists $\hat{x}^s + \hat{x}^c \in X_{\bar{m}}^s \oplus X_{\bar{m}}^c$ such that

$$(\tilde{P}_1^s(m) - \tilde{P}_2^s(m))(\bar{x}^s + \tilde{\ell}_{\bar{m}}^{cs}(\bar{x}^s, 0)) = \hat{x}^s + \hat{x}^c + \tilde{\ell}_{\bar{m}}^{cs}(\hat{x}^s, \hat{x}^c).$$

Define

$$\delta_s(\tilde{P}_1^s(m), \tilde{P}_2^s(m)) = \sup_{|\bar{x}^s| = 1} (|\hat{x}^s| + |\hat{x}^c|). \tag{11.32}$$

The next result is the analog to Lemma 11.7.

Lemma 11.13. *There exists a positive constant C such that*

$$\frac{1}{C}\|\tilde{P}_1^s(m) - \tilde{P}_2^s(m)\| \leq \delta_s(\tilde{P}_1^s(m), \tilde{P}_2^s(m)) \leq C\|\tilde{P}_1^s(m) - \tilde{P}_2^s(m)\|.$$

One may prove this lemma in the same fashion as Lemma 11.7. Set

$$d_s(\tilde{P}_1^s, \tilde{P}_2^s) = \sup_{m \in M} \delta_s(\tilde{P}_1^s(m), \tilde{P}_2^s(m)). \tag{11.33}$$

Then (11.33) defines an equivalent metric on Λ^s.

Proposition 11.14. *There exists $\sigma > 0$ such that if $\|\tilde{T} - T^t\|_1 \leq \sigma$, then $\mathcal{F}^s : \Lambda^s \to \Lambda^s$ is a contraction under the metric (11.33).*

Proof. For $\bar{x}^s \in X_{m_0}^s$ let

$$x^u + x^s + x^c = D\tilde{T}(m_0)(\bar{x}^s + \tilde{\ell}_{\bar{m}_0}^{cs}(\bar{x}^s, 0)). \tag{11.34}$$

The invariance of $\tilde{W}^{cs}(\epsilon)$ implies that $x^u = \tilde{\ell}_{\bar{m}_1}^{cs}(x^s, x^c)$. By Lemma 11.3, there exists a unique $\hat{x}^u + \hat{x}^s + \hat{x}^c \in X_{m_1}^u \oplus X_{m_1}^s \oplus X_{m_1}^c$ such that

$$x^u + x^s + x^c = \hat{x}^u + \hat{x}^c + \tilde{\ell}_{\bar{m}_1}^{cu}(\hat{x}^u, \hat{x}^c) + \hat{x}^s + \tilde{\ell}_{\bar{m}_1}^{cs}(\hat{x}^s, 0). \tag{11.35}$$

Note that $x^c = \hat{x}^c$. We claim that

$$|\hat{x}^s| \leq \left(\|DT^t(m_0)|_{X_{m_0}^s}\| + O(\epsilon) + C\sigma\right)|\bar{x}^s|. \tag{11.36}$$

To prove this claim, we first observe that

$$\hat{x}^u + \hat{x}^c + \tilde{\ell}_{\tilde{m}_1}^{cu}(\hat{x}^u, \hat{x}^c) \in T_{\tilde{m}_1}\tilde{M}.$$

Thus,

$$|\hat{x}^u| + |\tilde{\ell}_{\tilde{m}_1}^{cu}(\hat{x}^u, \hat{x}^c)| \le \frac{2\rho\mu}{1 - \rho\mu}\,|\hat{x}^c|. \tag{11.37}$$

Applying the projection $\Pi_{m_1}^c$ in (11.34), we find

$$
\begin{aligned}
|x^c| &= |\Pi_{m_1}^c D\tilde{T}(\tilde{m}_0)(\bar{x}^s + \tilde{\ell}_{\tilde{m}_0}^{cs}(\bar{x}^s, 0))| \\
&\le |\Pi_{m_1}^c (D\tilde{T}(\tilde{m}_0) - DT^t(m_0))(\bar{x}^s + \tilde{\ell}_{\tilde{m}_0}^{cs}(\bar{x}^s, 0))| \\
&\quad + \|\Pi_{m_1}^c - \Pi_{T^t(m_0)}^c\|\|DT^t(m_0)(\bar{x}^s + \tilde{\ell}_{\tilde{m}_0}^{cs}(\bar{x}^s, 0))| \\
&\le (O(\epsilon) + C\sigma)|\bar{x}^s|,
\end{aligned}
\tag{11.38}
$$

where (11.17) and (11.18) are used.

Similarly, applying the projection $\Pi_{m_1}^s$ to (11.34) and using (11.35) yields

$$\hat{x}^s + \tilde{\ell}_{\tilde{m}_1}^{cu}(\hat{x}^u, \hat{x}^c) = \Pi_{m_1}^s D\tilde{T}(\tilde{m}_0)(\bar{x}^s + \tilde{\ell}_{\tilde{m}_0}^{cs}(\bar{x}^s, 0)).$$

Hence, by (11.37)

$$
\begin{aligned}
|\hat{x}^s| &\le |\tilde{\ell}_{\tilde{m}_1}^{cu}(\hat{x}^s, \hat{x}^c)| + (O(\epsilon) + C\sigma)|\bar{x}^s| + |DT^t(m_0)\bar{x}^s| \\
&\le \frac{2\rho\mu}{1 - \rho\mu}\,|x^c| + (\|DT^t(m_0)|_{X_{m_1}^s}\| + O(\epsilon) + C\sigma)|\bar{x}^s|,
\end{aligned}
$$

which with (11.38) yields (11.36). This completes the proof of the claim. Let $\tilde{P}_1^s, \tilde{P}_2^s \in \Lambda^s$. For $i = 1, 2$, let

$$P_i^s(m_0) = (\mathcal{F}^s(\tilde{P}_i^s))(m_0).$$

Note that $\tilde{P}_1^s(m_1)(I - \tilde{P}_i^s(m_1)) = 0$ and $P_i^s(m_0)(I - P_i^s(m_0)) = 0$. Thus,

$$R(\tilde{P}_1^s(m_1) - \tilde{P}_2^s(m_1)) \subset T_{\tilde{m}_1}\tilde{M} \tag{11.39}$$

and

$$R(P_1^s(m_0) - P_2^s(m_0)) \subset T_{\tilde{m}_0}\tilde{M}. \tag{11.40}$$

For $\bar{x}^s \in X_{m_0}^s$ set

$$x_0^u + x_0^s + x_0^c = (P_1^s(m_0) - P_2^s(m_0))(\bar{x}^s + \tilde{\ell}_{\tilde{m}_0}^{cs}(\bar{x}^s, 0)).$$

By the definition of $\mathcal{F}^s$, we have

$$
\begin{aligned}
&x_0^u + x_0^s + x_0^s \\
&= (P_1^s(m_0) - P_2^s(m_0))(\bar{x}^s + \tilde{\ell}_{\tilde{m}_0}^{cs}(\bar{x}^s, 0)) \\
&= \left(D\tilde{T}(\tilde{m}_0)\big|_{T_{\tilde{m}_0}\tilde{W}^{cu}(\epsilon)} \right)^{-1} (\tilde{P}_1^s(m_1) - \tilde{P}_2^s(m_1)) D\tilde{T}(\tilde{m}_0)(\bar{x}^s + \tilde{\ell}_{\tilde{m}_0}^{cs}(\bar{x}^s, 0))
\end{aligned}
$$

Then, by the invariance, (11.40) yields that $x_0^u + x_0^s + x_0^c \in T_{\tilde{m}_0}\tilde{M}$.
Let

$$
x_1^u + x_1^s + x_1^c = D\tilde{T}(m_0)(x_0^u + x_0^s + x_0^c). \tag{11.41}
$$

Then we have that $x_1^u + x_1^s + x_1^c \in T_{\tilde{m}_1}\tilde{M}$. Thus,

$$
|x_0^u| + |x_0^s| \leq \frac{2\rho\mu}{1 - \rho\mu}\, |x_0^c|, \tag{11.42}
$$

$$
|x_0^s| + |x_0^c| \leq \frac{1}{1 - \rho\mu}\, |x_0^c|, \tag{11.43}
$$

and

$$
|x_1^u| + |x_1^s| \leq \frac{2\rho\mu}{1 - \rho\mu}\, |x_1^c|. \tag{11.44}
$$

Applying the projection $\Pi_{m_1}^c$ to (11.41) and using (11.17), (11.18) and (11.42), we obtain

$$
\begin{aligned}
|x_1^c| &= |\Pi_{m_1}^c D\tilde{T}(\tilde{m}_0)(x_0^u + x_0^s + x_0^c)| \\
&\geq \left(\inf\{|DT^t(m_0)x^c| : |x^c| = 1, x^c \in X_{m_0}^c\} - O(\epsilon) - C\sigma \right)|x_0^c|
\end{aligned} \tag{11.45}
$$

Hence, from (11.43)

$$
\begin{aligned}
|x_1^s| &+ |x_1^c| \\
&\geq \Big(\inf\{|DT^t(m_0)x^c| : |x^c| = 1, x^c \in X_{m_0}^c\} \\
&\quad - O(\epsilon) - C\sigma \Big)(1 - \rho\mu)(|x_0^s| + |x_0^c|).
\end{aligned} \tag{11.46}
$$

Observe from (11.34)

$$
x_1^u + x_1^s + x_1^c = (\tilde{P}_1^s(m_1) - \tilde{P}_2^s(m_1))(x^u + x^s + x^c) = (\tilde{P}_1^s(m_1) - \tilde{P}_2^s(m_1))(\hat{x}^s + \tilde{\ell}_{\tilde{m}_1}^{cs}(\hat{x}^s, 0)),
$$

which yields

$$
|x_1^s| + |x_1^c| \leq d_s(\tilde{P}_1^s, \tilde{P}_2^s)|\hat{x}^s|.
$$

Thus, using (11.36) and (11.46), we have

$$\frac{|x_0^s| + |x_0^c|}{|\bar{x}^s|} \leq \frac{\left(\left\|DT^t(m_0)\big|_{X_{m_0}^s}\right\| + O(\epsilon) + C\sigma\right)}{\left(\inf\{|DT^t(m_0)x^c| : |x^c| = 1, x^c \in X_{m_0}^c\} - O(\epsilon) - C\sigma\right)(1 - \rho\mu)} d_s(\tilde{P}_1^s, \tilde{P}_2^s),$$

which implies

$$d_s(P_1^s, P_2^s) \leq \frac{\left(\left\|DT^t(m_0)\big|_{X_{m_0}^s}\right\| + O(\epsilon) + C\sigma\right)}{\left(\inf\{|DT^t(m_0)x^c| : |x^c| = 1, x^c \in X_{m_0}^c\} - O(\epsilon) - C\sigma\right)(1 - \rho\mu)} d_s(\tilde{P}_1^s, \tilde{P}_2^s)$$

Note that (H3) implies

$$\frac{\left\|DT^t(m_0)\big|_{X_{m_0}^s}\right\|}{\inf\{|DT^t(m_0)x^c| : |x^c| = 1, x^c \in X_{m_0}^c\}} < \lambda.$$

Thus, by choosing μ, ϵ, and σ sufficiently small we have

$$d_s(\mathcal{F}^s(\tilde{P}_1^s), \mathcal{F}_1^s(\tilde{P}_2^s)) \leq \lambda_1 d_s(\tilde{P}_1^s, \tilde{P}_2^s).$$

This completes the proof. □

Note that Λ^s is complete under the metric (11.33) from Lemma 11.7. Since $\mathcal{F}^s$ is a contraction, by the contraction mapping theorem, we obtain

Proposition 11.15. *There exists $\sigma > 0$ such that if $\|\tilde{T} - T^t\|_1 \leq \sigma$, there exists $\tilde{\Pi}^s \in \Lambda^s$ such that*

$$\tilde{\Pi}^s(K^{-1}(\tilde{m}_1))D\tilde{T}(\tilde{m}_0) = D\tilde{T}(\tilde{m}_0)\tilde{\Pi}^s(K^{-1}(\tilde{m}_0)).$$

Proposition 11.16. $\|\Pi^s - \tilde{\Pi}^s\|_0 \to 0$ *as* $\|\tilde{T} - T^t\|_1 \to 0$.

Proof. The idea of the proof is the same as that for Proposition 11.10. We first notice that $\mathcal{N}^s\omega \in \Lambda^s$ from the proof of Lemma 11.3. The first step is to estimate

$$d_s(\mathcal{N}^s\omega, \mathcal{F}^s(\mathcal{N}^s\omega)).$$

For each $\tilde{x}^u + \tilde{x}^s + \tilde{x}^c \in X_{m_0}^u \oplus X_{m_0}^s \oplus X_{m_0}^c$, by Lemma 11.3, there exists a unique $\bar{x}^u + \bar{x}^s + \bar{x}^c \in X_{m_0}^u \oplus X_{m_0}^s \oplus X_{m_0}^c$ such that

$$\tilde{x}^u + \tilde{x}^s + \tilde{x}^c = \bar{x}^u + \bar{x}^c + \tilde{\ell}_{\tilde{m}_0}^{cu}(\bar{x}^u, \bar{x}^c) + \bar{x}^s + \tilde{\ell}_{\tilde{m}_1}^{cs}(\bar{x}^s, 0).$$

Note that $\tilde{x}^c = \bar{x}^c$. Write

$$D\tilde{T}(m_0)(\bar{x}^s + \tilde{\ell}^{cs}_{\tilde{m}_0}(\bar{x}^s, 0))) = \hat{x}^u + \hat{x}^c + \tilde{\ell}^{cu}_{\tilde{m}_1}(\hat{x}^u, \hat{x}^c) + \hat{x}^s + \tilde{\ell}^{cs}_{\tilde{m}_1}(\hat{x}^s, 0). \qquad (11.47)$$

Let

$$x_0^u + x_0^s + x_0^c = \Big(\mathcal{N}^s(m_0)\omega(m_0) - (\mathcal{F}^s(\mathcal{N}^s\omega))(m_0) \Big)(\bar{x}^s + \tilde{\ell}^{cs}_{\tilde{m}_0}(\bar{x}^s, 0)).$$

Clearly, $x_0^u = \tilde{\ell}^{cs}_{\tilde{m}_0}(x_0^s, x_0^c)$. On the other hand, by the definition of $\mathcal{F}^s$

$$\Big(\mathcal{N}^s(m_0)\omega(m_0) - (\mathcal{F}^s(\mathcal{N}^s\omega))(m_0) \Big)(\bar{x}^s + \tilde{\ell}^{cs}_{\tilde{m}_0}(\bar{x}^s, 0))$$
$$= \Big(D\tilde{T}(\tilde{m}_0)\big|_{T_{\tilde{m}_0}\tilde{W}^{cu}(\epsilon)} \Big)^{-1}(I - \mathcal{N}^s(m_1)\omega(m_1))D\tilde{T}(\tilde{m}_0)(\bar{x}^s + \tilde{\ell}^{cs}_{\tilde{m}_1}(\bar{x}^s, 0)),$$

which is in $T_{\tilde{m}_0}\tilde{W}^{cu}(\epsilon)$. Hence,

$$x_0^u + x_0^s + x_0^c \in T_{\tilde{m}_0}\tilde{M}.$$

Thus,

$$|x_0^u| + |x_0^s| \leq \frac{2\rho\mu}{1 - \rho\mu} \, |x_0^c| \qquad (11.48)$$

and

$$|x_0^s| + |x_0^c| \leq \frac{1}{1 - \rho\mu} \, |x_0^c|. \qquad (11.49)$$

Computing $D\tilde{T}(\tilde{m}_0)(x_0^u + x_0^s + x_0^c)$, we find

$$D\tilde{T}(\tilde{m}_0)(x_0^u + x_0^s + x_0^c) = \hat{x}^u + \hat{x}^c + \tilde{\ell}^{cu}_{\tilde{m}_1}(\hat{x}^u, \hat{x}^c). \qquad (11.50)$$

From the invariance, we have that $\hat{x}^u + \hat{x}^c + \tilde{\ell}^{cu}_{\tilde{m}_1}(\hat{x}^u, \hat{x}^c) \in T_{\tilde{m}_1}\tilde{W}^{cu}(\epsilon)$, and hence

$$\hat{x}^u + \hat{x}^c + \tilde{\ell}^{cu}_{\tilde{m}_1}(\hat{x}^u, \hat{x}^c) \in T_{\tilde{m}_1}\tilde{M}.$$

Again,

$$|\hat{x}^u| + |\tilde{\ell}^{cu}_{\tilde{m}_1}(\hat{x}^u, \hat{x}^c)| \leq \frac{2\rho\mu}{1 - \rho\mu} \, |\hat{x}^c|.$$

Applying the projection $\Pi^c_{m_1}$ to (11.50) and using (11.17), (11.18) and (11.48), we obtain

$$|\hat{x}^c| = |\Pi^c_{m_1} D\tilde{T}(\tilde{m}_0)(x_0^u + x_0^s + x_0^c)|$$
$$\geq \Big(\inf\{|DT^t(m_0)x^c| : |x^c| = 1, x^c \in X^c_{m_0}\} - O(\epsilon) - C\sigma \Big)|x_0^c| \qquad (11.51)$$

Combining (11.47) with (11.34) and (11.35) and using (11.38), (11.49) and (11.51), we have

$$|x_0^s| + |x_0^c| \leq (O(\epsilon) + C\sigma)|\bar{x}^s|$$

which implies that

$$\delta_s(\mathcal{N}^s(m_0)\omega(m_0), (\mathcal{F}^s(\mathcal{N}^s\omega))(m_0)) \leq O(\epsilon) + C\sigma$$

Since $\tilde{\Pi}^s$ is the fixed point of the mapping $\mathcal{F}^s$ which has Lipschitz constant $\lambda_1 < 1$, a simple computation yields

$$\|\mathcal{N}^s\omega - \tilde{\Pi}^s\|_s \leq (\sum_{k=0}^{\infty} \lambda_1^k)(O(\epsilon) + C\sigma) = O(\epsilon) + C\sigma \qquad (11.52)$$

with a different constant C. Next, we estimate $\|\mathcal{N}^s\omega - \Pi^s\|$. From the definition of $\tilde{\psi}^s(m)$, it is easy to see

$$\|(\tilde{\psi}^s(m_0))^{-1} - I\| \leq \frac{\rho\mu}{1 - \rho\mu}, \qquad (11.53)$$

which implies

$$|x^s| \leq \frac{1}{1 - \rho\mu}(|\tilde{x}^s| + |\tilde{x}^u| + |\tilde{x}^c|).$$

Thus,

$$\begin{aligned}
|(\mathcal{N}^s(m_0)\omega(m_0) &- \Pi^s_{m_0})(\tilde{x}^u + \tilde{x}^s + \tilde{x}^c)| \\
&= |\bar{x}^s + \tilde{\ell}^{cs}_{\tilde{m}_0}(\bar{x}^s, 0) - \tilde{x}^s| \\
&\leq |\bar{x}^s - \tilde{x}^s| + \rho\mu|\hat{x}^s| \\
&\leq \frac{2\rho\mu}{1 - \rho\mu}(|\tilde{x}^u| + |\tilde{x}^s| + |\tilde{x}^c|) \\
&\leq C\frac{2\rho\mu}{1 - \rho\mu}|\tilde{x}^u + \tilde{x}^c + \tilde{x}^s|,
\end{aligned}$$

which implies

$$\|\mathcal{N}^s\omega - \Pi^s\| \leq C\frac{2\rho\mu}{1 - \rho\mu}.$$

By Lemma 11.13, we obtain

$$\begin{aligned}
\|\tilde{\Pi}^s - \Pi^s\| &\leq \|\tilde{\Pi}^s - \mathcal{N}^s\omega\| + \|\mathcal{N}^s\omega - \Pi^s\| \\
&\leq C d_s(\tilde{\Pi}^s, \mathcal{N}^s\omega) + \|\mathcal{N}^s\omega - \Pi^s\| \\
&\leq O(\epsilon) + C\sigma + \frac{2\rho\mu C}{1 - \rho\mu}.
\end{aligned}$$

For any $\mathcal{E} > 0$, choosing μ, ϵ^* and σ sufficiently small, we have that if $\|\tilde{T} - T^t\|_1 < \sigma$, then

$$\|\tilde{\Pi}^s - \Pi^s\| < \mathcal{E}.$$

This completes the proof. $\square$

Let

$$\tilde{\Pi}^c = I - \tilde{\Pi}^u - \tilde{\Pi}^s.$$

It is easy to see

Proposition 11.17.

    (i) $\tilde{\Pi}^c \in C^0(M, L(X, X))$

    (ii) $R(\Pi^c(m)) = T_{\tilde{m}}\tilde{M}$ and $(\Pi^c(m))^2 = \Pi^c(m)$.

    (iii) $\|\tilde{\Pi}^c - \Pi^c\|_0 \to 0$ as $\|\tilde{T} - T^t\|_1 \to 0$.

Proof of Theorem 11.1. For $\alpha = u, s, c$, let $\tilde{\Pi}^\alpha_{\tilde{m}} = \tilde{\Pi}^\alpha(K^{-1}(\tilde{m}))$ and $\tilde{X}^\alpha_{\tilde{m}} = R(\tilde{\Pi}^\alpha_{\tilde{m}})$. Clearly, for each $\tilde{m} \in \tilde{M}$ we have a decomposition

$$X = \tilde{X}^u_{\tilde{m}} \oplus X^s_{\tilde{m}} \oplus X^c_{\tilde{m}}$$

of closed subspaces with $\tilde{X}^c_{\tilde{m}}$ the tangent space to $\tilde{M}$ at $\tilde{m}$. Propositions 11.9, 11.15, and 11.17 yield

$$D\tilde{T}(\tilde{m}) : \tilde{X}^\alpha_{\tilde{m}} \to \tilde{X}^\alpha_{\tilde{T}(\tilde{m})} \text{ for } \alpha = c, u, s.$$

Furthermore, $D\tilde{T}(\tilde{m})$ is an isomorphism from $\tilde{X}^u_{\tilde{m}}$ onto $\tilde{X}^u_{\tilde{T}(\tilde{m})}$. Next we claim there exists $\sigma > 0$ such that if $\|\tilde{T} - T^t\|_1 \leq \sigma$ then

$$\lambda_1 \inf \left\{ \left| D\tilde{T}(\tilde{m})\tilde{x}^u \right| : \tilde{x}^u \in \tilde{X}^u_{\tilde{m}}, |\tilde{x}^u| = 1 \right\} > \max \left\{ 1, \|D\tilde{T}(\tilde{m})\big|_{\tilde{X}^c_{\tilde{m}}}\| \right\}$$

$$\lambda_1 \min \left\{ 1, \inf \left\{ \left| D\tilde{T}(\tilde{m})\tilde{x}^c \right| : \tilde{x}^c \in \tilde{X}^c_{\tilde{m}}, |\tilde{x}^c| = 1 \right\} \right\} > \|D\tilde{T}(\tilde{m})\big|_{\tilde{X}^s_{\tilde{m}}}\|$$

Let us first look at

$$\lambda_1 \inf \left\{ \left| D\tilde{T}(\tilde{m})\tilde{x}^u \right| : \tilde{x}^u \in \tilde{X}^u_{\tilde{m}}, |\tilde{x}^u| = 1 \right\} > 1. \tag{11.54}$$

From (H3), we have

$$\zeta = \inf_{|x^u|=1} |DT^t(m)x^u| > \frac{1}{\lambda},$$

and so

$$|DT^t(m)x^u| \geq \zeta|x^u|.$$

Thus,

$$
\begin{aligned}
|D\tilde{T}(\tilde{m})\tilde{x}^u| &= |DT^t(m)\Pi_m\tilde{x}^u| - |D\tilde{T}(\tilde{m})\tilde{x}^u - DT^t(m)\Pi_m\tilde{x}^u| \\
&\geq \zeta|\Pi_m^u\tilde{x}^u| - \|D\tilde{T}(\tilde{m}) - DT^t(m)\| \ |\tilde{x}^u| \\
&\quad - \|DT^t(m)\| \ \|\Pi_{\tilde{m}}^u - \Pi_m^u\| \ |\tilde{x}^u| \\
&\geq (\zeta - C\|\tilde{\Pi}_{\tilde{m}}^u - \Pi_m^u\| - O(\epsilon) - C\sigma)|\tilde{x}^u|,
\end{aligned}
$$

where (11.17) and (11.18) are used.

Using Proposition 11.10, we may choose ϵ^* and σ sufficiently small so that

$$
\zeta - C\|\tilde{\Pi}_{\tilde{m}}^u - \Pi_m^u\| - O(\epsilon) - C\sigma > \frac{1}{\lambda_1},
$$

giving (11.54). One may prove the other estimates in the same fashion by using (H3) and Propositions 11.10, 11.16 and 11.17. The proof is complete □

Another direct consequence of Propositions 11.9, 11.15 and 11.17 is

Proposition 11.18. *For each $\tilde{m} \in \tilde{M}$*

$$
\begin{aligned}
T_{\tilde{m}}\tilde{W}^{cs}(\epsilon) &= \tilde{X}_{\tilde{m}}^s \oplus \tilde{X}_{\tilde{m}}^c, \\
T_{\tilde{m}}\tilde{W}^{cu}(\epsilon) &= \tilde{X}_{\tilde{m}}^u \oplus \tilde{X}_{\tilde{m}}^c.
\end{aligned}
$$

We now point out how to obtain Theorems A–D, stated in Section 3. Theorem A is extracted from Theorem 6.3, Proposition 6.10, Proposition 6.12, and Theorem 9.1; Theorem B is extracted from Theorem 7.3, Proposition 7.10, and Theorem 8.1; Theorem C is extracted from Theorem 10.1, Proposition 10.2, Theorem 11.1, and Proposition 11.18; and finally, Theorem D follows from Proposition 6.10, Proposition 7.10, and Theorem 10.3.

12. Invariant Manifolds for Perturbed Semiflow.

In the previous sections, we obtained the existence of a compact, normally hyperbolic invariant manifold, $\tilde{M}$, for the map $\tilde{T}$, a C^1 perturbation of the time t-map T^t, where $t > t_0$. In doing so, we also obtained the stable and unstable manifolds of $\tilde{M}$ under the map $\tilde{T}$. We now consider the perturbed semiflow $\tilde{T}^t$ of T^t. The argument we use is different from that typically used in the finite dimensional case since we do not have a vector field, nor do we cut-off the semiflow outside the tubular neighborhood.

Assume that the perturbed semiflow $\tilde{T}^t$ is continuous on $[0, \infty) \times X$ into X, and that for each $t \geq 0, \tilde{T}^t : X \to X$ is C^1.

Let us first recall some notation from Section 3. We use B to denote a fixed neighborhood of M in X containing the tubular neighborhood $\Theta(X^u(\epsilon_0)) \oplus X^s(\epsilon_0))$. We defined

$$\|\tilde{T}^t - T^t\|_0 \equiv \sup_{x \in B} |\tilde{T}^t(x) - T^t(x)|$$

and

$$\|\tilde{T}^t - T^t\|_1 \equiv \|\tilde{T}^t - T^t\|_0 + \sup_{x \in B} \|D\tilde{T}^t(x) - DT^t(x)\|.$$

Let $t_1 > t_0$ be fixed. By Theorem 9.1 and Proposition 6.10, there exists $\epsilon^* > 0$ such that for each $\epsilon < \epsilon^*$ there is a $\sigma > 0$ such that if $\|\tilde{T}^{t_1} - T^{t_1}\|_1 < \sigma$ then $\tilde{T}^{t_1}$ has a unique C^1 center-unstable manifold $\tilde{W}^{cu}(\epsilon)$ in the tubular neighborhood $\Theta(X^u(\epsilon) \oplus X^s(\epsilon))$ which satisfies

$$\tilde{T}^{t_1}(\tilde{W}^{cu}(\epsilon)) \cap \Theta(X^u(\epsilon) \oplus X^s(\epsilon)) = \tilde{W}^{cu}(\epsilon).$$

Proposition 12.1. *There exists $\epsilon^* > 0$ such that for each $\epsilon < \epsilon^*$ there are $\sigma > 0$ and $\epsilon' < \epsilon$ such that if*

$$\|\tilde{T}^{t_1} - T^{t_1}\|_1 < \sigma \text{ and } \|\tilde{T}^t - T^t\|_0 < \sigma, \text{ for } 0 \leq t \leq t_1,$$

then for all $t \in [0, t_1]$

$$\tilde{T}^t(\tilde{W}^{cu}(\epsilon')) \subset \tilde{W}^{cu}(\epsilon)$$

Proof. For the fixed $t_1 > t_0$, by Theorem 9.1, there exist positive constants $\tilde{\epsilon}^*$, $\epsilon^* = \epsilon^*(\tilde{\epsilon})$, $\delta^* = \delta^*(\epsilon) < \epsilon$ and $\sigma = \sigma(\epsilon, \delta)$ such that if $\tilde{\epsilon} < \tilde{\epsilon}^*$, $\epsilon < \epsilon^*$, $\delta < \delta^*$, and $\tilde{T}^t$ satisfies $\|\tilde{T}^{t_1} - T^{t_1}\|_1 < \sigma$, then $\tilde{T}^{t_1}$ has a C^1 center-unstable manifold with Lipschitz constant $\rho\mu$,

$$\tilde{W}^{cu}(\epsilon) = \Theta(\text{gr}(\tilde{h}^{cu})),$$

where $\tilde{h}^{cu} \in \Gamma^{cu}(\epsilon, \mu, \delta, \tilde{\epsilon})$. Let $\epsilon < \epsilon^*$ be fixed. Then, by Lemma 5.1, there exists $\epsilon' \leq \epsilon$ such that for $t \in [0, t_1]$

$$T^t(\Theta(X^u(\epsilon') \oplus X^s(\epsilon'))) \subset \Theta(X^u(\tfrac{1}{2}\epsilon) \oplus X^s(\tfrac{1}{2}\epsilon)).$$

By choosing σ sufficiently small, we have that if $\|\tilde{T}^t - T^t\|_0 < \sigma$, for $0 \leq t \leq t_1$, then for each $t \in [0, t_1]$

$$\tilde{T}^t(\Theta(X^u(\epsilon') \oplus X^s(\epsilon'))) \subset \Theta(X^u(\epsilon) \oplus X^s(\epsilon)). \qquad (12.1)$$

We also choose σ sufficiently small such that both $\tilde{W}^{cu}(\epsilon')$ and $\tilde{W}^{cu}(\epsilon)$ exist. We denote the corresponding $\tilde{h}^{cu}$ for ϵ' and ϵ by $\tilde{h}_{\epsilon'}^{cu}$ and $\tilde{h}_{\epsilon}^{cu}$ respectively. Proposition 6.10 implies that $\tilde{h}_{\epsilon'}^{cu} = \tilde{h}_{\epsilon}^{cu}$ on $X^u(\epsilon')$. We want to show that $\tilde{T}^t(\tilde{W}^{cu}(\epsilon')) \subset \tilde{W}^{cu}(\epsilon)$ for $t \in [0, t_1]$.

For each $(m_0, x_0^u) \in X^u(\epsilon')$, from the fact that $\tilde{T}^{t_1}(\tilde{W}^{cu}(\epsilon')) \cap (X^u(\epsilon') \oplus X^s(\epsilon')) = \tilde{W}^{cu}(\epsilon')$, there is a sequence of points $(m_k, x_k^u) \in X^u(\epsilon')$ such that for $k \geq 0$

$$\tilde{T}^{t_1}(m_{k+1} + x_{k+1}^u + \tilde{h}_\epsilon^{cu}(m_{k+1}, x_{k+1}^u)) = m_k + x_k^u + \tilde{h}_\epsilon^{cu}(m_k, x_k^u).$$

Let $x_k^s = \tilde{h}_\epsilon^{cu}(m_k, x_k^u)$ and $m_k(t) + x_k^u(t) + x_k^s(t) = \tilde{T}^t(m_k + x_k^u + x_k^s)$ for $k \geq 0$. From (12.1), $(m_k(t), x_k^u(t) + x_k^s(t)) \in X^u(\epsilon) \oplus X^s(\epsilon)$. Observe that

$$\tilde{T}^{t_1}(m_{k+1}(t) + x_{k+1}^u(t) + x_{k+1}^s(t)) = m_k(t) + x_k^u(t) + x_k^s(t)$$

Thus, applying Lemma 6.8, we obtain for all $k \geq 0$

$$|x_k^s(t) - \tilde{h}_\epsilon^{cu}(m_k(t), x_k^u(t))| \leq \lambda_1 |x_{k+1}^s(t) - \tilde{h}_\epsilon^{cu}(m_{k+1}(t), x_{k+1}^u(t))| \qquad (12.2)$$

Note that $\lambda_1 < 1$ and $|x_k^\alpha(t)| \leq \epsilon$ for $\alpha = u, s$. Hence, (12.2) yields

$$x_0^s(t) = \tilde{h}_\epsilon^{cu}(m_0(t), x_0^u(t)),$$

In other words, $\tilde{T}^t(m_0 + x_0^u + \tilde{h}_{\epsilon'}^{cu}(m_0, x_0^u)) \in \tilde{W}^{cu}(\epsilon)$, for $0 \leq t \leq t_1$, which completes the proof. $\square$

Theorem 12.2. *There exits ϵ' and σ such that if*

$$\|\tilde{T}^{t_1} - T^{t_1}\|_1 < \sigma \text{ and } \|\tilde{T}^t - T^t\|_0 < \sigma, \text{ for } 0 \leq t \leq t_1,$$

then

(i) $\tilde{T}^t(\tilde{W}^{cu}(\epsilon')) \cap \Theta(X^u(\epsilon') \oplus X^s(\epsilon')) \subset \tilde{W}^{cu}(\epsilon')$ *for* $0 \leq t \leq t_1$;

(ii) *For* $(m_0, x_0^u) \in X^u(\epsilon')$, *if* $\tilde{T}^t(m_0 + x_0^u + \tilde{h}^{cu}(m_0, x_0^u)) \in X^u(\epsilon') \oplus X^s(\epsilon')$ *for* $0 \leq t \leq t_2$, *then* $\tilde{T}^t(m_0 + x_0^u + \tilde{h}^{cu}(m_0, x_0^u)) \in \tilde{W}^{cu}(\epsilon')$ *for* $t \in [0, t_2]$;

(iii) $\tilde{W}^{cu}(\epsilon') \supset \cap_{t=0}^\infty \tilde{\mathcal{A}}_t$, *where*

$$\tilde{\mathcal{A}}_t = \{(m_0, x_0^u, x_0^s) \in X^u(\epsilon') \oplus X^s(\epsilon') : \exists\, (m_1, x_1^u, x_1^s) \in X^u(\epsilon') \oplus X^s(\epsilon')$$

$$\text{such that } \tilde{T}^t(m_1 + x_1^u + x_1^s) = m_0 + x_0^u + x_0^s,$$

$$\text{and } \tilde{T}^\tau(m_1 + x_1^u + x_1^s) \in X^u(\epsilon') \oplus X^s(\epsilon'), \text{ for all } 0 \leq \tau \leq t\}$$

Proof. (i)–(ii) follows directly from Proposition 12.1. Note that $\tilde{\mathcal{A}}_{kt_1} \subset \mathcal{A}_k$, where $\mathcal{A}_k$ is given in Proposition 6.10. Hence (iii) holds. ☐

From Theorem 8.1 and Proposition 7.10, there exists $\epsilon^* > 0$ such that for each $\epsilon < \epsilon^*$ there is a $\sigma > 0$ such that if $\|\tilde{T}^{t_1} - T^{t_1}\|_1 < \sigma$ then T^{t_1} has a unique C^1 center-stable manifold $\tilde{W}^{cs}(\epsilon)$ in the tubular neighborhood $\Theta(X^u(\epsilon) \oplus X^s(\epsilon))$ which satisfies

$$\tilde{T}^{t_1}(\tilde{W}^{cs}(\epsilon)) \subset \tilde{W}^{cs}(\epsilon).$$

Proposition 12.3. *There exists $\epsilon^* > 0$ such that for each $\epsilon < \epsilon^*$ there are $\sigma > 0$ and $\epsilon' < \epsilon$ such that if*

$$\|\tilde{T}^{t_1} - T^{t_1}\|_1 < \sigma \text{ and } \|\tilde{T}^t - T^t\|_0 < \sigma, \text{ for } 0 \leq t \leq t_1,$$

then

$$\tilde{T}^t(\tilde{W}^{cs}(\epsilon')) \subset \tilde{W}^{cs}(\epsilon), \text{ for } 0 \leq t \leq t_1.$$

Proof. One may establish this proposition in the same fashion as Proposition 12.1. The only difference is that we define the sequence (m_k, x_k^u) by

$$\tilde{T}^{t_1}(m_{k-1} + x_{k-1}^u + \tilde{h}_{\epsilon'}^{cs}(m_{k-1}, x_{k-1}^u)) = m_k + x_k^u + \tilde{h}_{\epsilon'}^{cs}(m_k, x_k^u).$$

We omit the details. ☐

Theorem 12.4. *There exist ϵ' and σ such that if*

$$\|\tilde{T}^{t_1} - T^{t_1}\|_1 < \sigma \text{ and } \|\tilde{T}^t - T^t\|_0 < \sigma, \text{ for } 0 \leq t \leq t_1,$$

then

(i) *For all $t \geq 0$, $\tilde{T}^t(\tilde{W}^{cs}(\epsilon')) \cap \Theta(X^u(\epsilon') \oplus X^s(\epsilon')) \subset W^{cs}(\epsilon')$*
    and
(ii) *$\{x : \tilde{T}^t(x) \in \Theta(X^u(\epsilon') \oplus X^s(\epsilon')), t \geq 0\} \subset \tilde{W}^{cs}(\epsilon')$.*

Proof. (i) is a direct consequence of Proposition 12.3. (ii) follows from Proposition 7.10. ☐

From Theorem 10.1, we have that $\tilde{M} = \tilde{W}^{cs}(\epsilon) \cap \tilde{W}^{cu}(\epsilon)$ is a C^1 compact invariant manifold for the time-t_1 map $\tilde{T}^{t_1}$. The next result gives that $\tilde{M}$ is invariant for the semiflow $\tilde{T}^t$.

Proposition 12.5. *There exists $\epsilon^* > 0$ such that for each $\epsilon < \epsilon^*$ there is $\sigma > 0$ such that if*

$$\|\tilde{T}^{t_1} - T^{t_1}\|_1 < \sigma \text{ and } \|\tilde{T}^t - T^t\|_0 < \sigma, \text{ for } 0 \leq t \leq t_1,$$

then $\tilde{T}^t$ has a unique C^1 compact invariant manifold $\tilde{M}$ in the tubular neighborhood $\Theta(X^u(\epsilon) \oplus X^s(\epsilon))$ and for each $t \geq 0$, $\tilde{T}^t$ is a C^1 diffeomorphism from $\tilde{M}$ onto $\tilde{M}$.

Proof. Let σ and ϵ' satisfy the requirements of Proposition 12.1 and Proposition 12.3. Since $\tilde{T}^{t_1}$ has a unique compact connected invariant manifold in $\Theta(X^u(\epsilon) \oplus X^s(\epsilon))$, we have

$$\tilde{W}^{cs}(\epsilon') \cap \tilde{W}^{cu}(\epsilon') = \tilde{W}^{cs}(\epsilon) \cap \tilde{W}^{cu}(\epsilon).$$

Propositions 12.1 and 12.3 imply that, for $0 \leq t \leq t_1$,

$$\tilde{T}^t(\tilde{M}) \subset \tilde{M}.$$

Note that each $t > t_1$ may be written as $t = kt_1 + \tilde{t}$ for a positive integer k and $0 \leq \tilde{t} \leq t_1$. Hence, $\tilde{T}^t = \tilde{T}^{kt_1 + \tilde{t}} = \tilde{T}^{kt_1}\tilde{T}^{\tilde{t}}$ which implies that $T^t(\tilde{M}) \subset \tilde{M}$. For each $0 \leq t \leq t_1$, write $\tilde{T}^{t_1} = \tilde{T}^{t_1 - t}T^t$. Then, it follows that $\tilde{T}^t$ is a C^1 diffeomorphism from the fact that $\tilde{T}^{t_1}$ is a diffeomorphism from $\tilde{M}$ onto $\tilde{M}$, which yields that $\tilde{T}^t$ is a C^1 diffeomorphism for all $t \geq 0$. This completes the proof. $\square$

As a corollary of Theorem 10.3, by using Propositions 12.1 and 12.3, we obtain

Theorem 12.6.

    (i) $\lim_{t \to \infty} d(\tilde{T}^t(x), \tilde{M}) = 0$, *uniformly for* $x \in \tilde{W}^{cs}(\epsilon')$.
    and
    (ii) $\lim_{t \to \infty} d(\tilde{T}^{-t}(x), \tilde{M}) = 0$, *uniformly for* $x \in \tilde{W}^{cu}(\epsilon')$.

Theorem 12.7. $\tilde{M}$ *is a normally hyperbolic manifold.*

Proof. We first show that $\tilde{X}^\alpha$ for $\alpha = u, s, c$, defined in Section 11, are invariant for the semiflow $\tilde{T}$ provided that σ is sufficiently small. Proposition 12.5 and Proposition 11.17 imply that $\tilde{X}^c$ is invariant. We consider $\tilde{X}^s$. For $\tilde{m}_0 \in \tilde{M}$ and $\tilde{x}^s \in \tilde{X}^s_{\tilde{m}_0}$, it is enough to show that $D\tilde{T}^t(\tilde{m}_0)\tilde{x}^s \in \tilde{X}^s_{\tilde{T}^t(\tilde{m}_0)}$ for $0 \leq t \leq t_1$. Write

$$D\tilde{T}^t(\tilde{m}_0)\tilde{x}^s = \tilde{x}^u(t) + \tilde{x}^s(t) + \tilde{x}^c(t), \qquad (12.3)$$

where $\tilde{x}^\alpha(t) \in \tilde{X}^\alpha_{\tilde{T}^t(\tilde{m}_0)}$. Proposition 11.15 and the invariance of $\tilde{W}^{cs}(\epsilon)$ implies that $\tilde{x}^u(t) = 0$. We claim that $\tilde{x}^c(t) = 0$, for $0 \leq t \leq t_1$. Suppose that for some $\bar{t} \in [0, t_1]$, $\tilde{x}^c(\bar{t}) \neq 0$. Let $\tilde{m}(t) = \tilde{T}^t(\tilde{m}_0)$ and $m(t) = K^{-1}\tilde{m}(t)$. For simplicity, we set

$$\tilde{x}^c(kt_1 + \bar{t}) = D\tilde{T}^{kt_1}(\tilde{m}(\bar{t}))\tilde{x}^c(\bar{t}),$$
$$\tilde{x}^s(kt_1 + \bar{t}) = D\tilde{T}^{kt_1}(\tilde{m}(\bar{t}))\tilde{x}^s(\bar{t}) \qquad (12.4)$$

Note that for $\alpha = s, c$

$$\tilde{x}^\alpha((k+1)t_1 + \bar{t}) = D\tilde{T}^{t_1}(\tilde{m}(kt_1 + \bar{t}))\tilde{x}^\alpha(kt_1 + \bar{t}).$$

Thus, from Theorem 11.1 it follows that

$$|\tilde{x}^c((k-1)t_1 + \bar{t})||\tilde{x}^s(kt_1 + \bar{t})| \leq \lambda_1 |\tilde{x}^s((k-1)t_1 + \bar{t})||\tilde{x}^c(kt_1 + \bar{t})|,$$

which implies

$$|\tilde{x}^s(kt_1 + \bar{t})| \leq \lambda_1^k \frac{|\tilde{x}^s(\bar{t})|}{|\tilde{x}^c(\bar{t})|} |\tilde{x}^c(kt_1 + \bar{t})|, \tag{12.5}$$

where the assumption that $\tilde{x}^c(\bar{t}) \neq 0$ is used. Thus, using (12.3) and (12.4), we have

$$\begin{aligned}
D\tilde{T}^{(k+1)t_1}(\tilde{m}_0)\tilde{x}^s \\
= D\tilde{T}^{t_1 - \bar{t}}(\tilde{m}(kt_1 + \bar{t}))D\tilde{T}^{kt_1 + \bar{t}}(\tilde{m}_0)\tilde{x}^s \\
= D\tilde{T}^{t_1 - \bar{t}}(\tilde{m}(kt_1 + \bar{t}))(\tilde{x}^s(kt_1 + \bar{t}) + \tilde{x}^c(kt_1 + \bar{t})).
\end{aligned} \tag{12.6}$$

Applying the projection $\tilde{\Pi}^c_{\tilde{m}((k+1)t_1)}$ to (12.6), we obtain

$$0 = \tilde{\Pi}^c_{\tilde{m}((k+1)t_1)}\Big(D\tilde{T}^{t_1 - \bar{t}}(\tilde{m}(kt_1 + \bar{t}))(\tilde{x}^s(kt_1 + \bar{t}) + \tilde{x}^c(kt_1 + \bar{t}))\Big).$$

Using (12.5), we obtain

$$\begin{aligned}
0 \geq |D\tilde{T}^{t_1 - \bar{t}}(\tilde{m}(kt_1 + \bar{t}))\tilde{x}^c(kt_1 + \bar{t})| \\
- C\|D\tilde{T}^{t_1 - \bar{t}}\|_{\tilde{M}} \lambda_1^k \frac{|\tilde{x}^s(\bar{t})|}{|\tilde{x}^c(\bar{t})|} |\tilde{x}^c(kt_1 + \bar{t})|,
\end{aligned} \tag{12.7}$$

where $\| \cdot \|_{\tilde{M}}$ is the supremum over $\tilde{M}$ of the norms of the operators. Since $\tilde{T}^{t_1 - \bar{t}}$ is a diffeomorphism on $\tilde{M}$ and $\tilde{M}$ is compact, this inequality is impossible for k large enough. Therefore $\tilde{X}^s$ is invariant for $\tilde{T}^t$. Similarly, one may prove that $\tilde{X}^u$ is invariant.

To prove $\tilde{M}$ is a normally hyperbolic invariant manifold for $\tilde{T}^t$, first we notice that Lemma 2.2 can be applied to $\tilde{T}^t$ and $\tilde{M}$ without any modification. We also need the following lemma.

Lemma 12.8. *There exists $\bar{t}_0 > 0$ such that for $t_4 > t_3 > \bar{t}_0$ there is a constant $C > 0$ such that $\|D\tilde{T}^t(\tilde{m})\| \leq C$ for all $\tilde{m} \in \tilde{M}$ and $t \in [t_3, t_4]$.*

The proof of this lemma will be given after we complete the proof Theorem 12.7. We want to show that for sufficiently large t,

$$\lambda_1 \inf \Big\{ |D\tilde{T}^t(\tilde{m})\tilde{x}^u| \; : \; \tilde{x}^u \in \tilde{X}^u_{\tilde{m}}, |\tilde{x}^u| = 1 \Big\} > \max \Big\{ 1, \|D\tilde{T}^t(\tilde{m})|_{\tilde{X}^c_{\tilde{m}}}\| \Big\} \tag{12.8}$$

and

$$\lambda_1 \min \Big\{ 1, \inf \Big\{ |D\tilde{T}^t(\tilde{m})\tilde{x}^c| \; : \; \tilde{x}^c \in \tilde{X}^c_{\tilde{m}}, |\tilde{x}^c| = 1 \Big\} \Big\} > \|D\tilde{T}^t(\tilde{m})|_{\tilde{X}^s_{\tilde{m}}}\| \tag{12.9}$$

From Section 11, (12.8), and (12.9) hold for $t = t_1$. It follows from Lemma 12.8 that there exist a positive integer k_0 and a constant $C > 0$ such that $\|D\tilde{T}^t(\tilde{m})\| < C$ for all $t \in [k_0 t_1, (k_0+1)t_1]$ and $\tilde{m} \in \tilde{M}$. Hence, for $t = kt_1 + \tilde{t}$, where $\tilde{t} \in [k_0 t_1, (k_0+1)t_1]$ and k is a positive integer, we have for all $\tilde{m} \in \tilde{M}$,

$$\|D\tilde{T}^t(\tilde{m})|_{\tilde{X}^s_{\tilde{m}}}\| < C\lambda_1^k. \tag{12.10}$$

Note that $\lambda_1 \in (\lambda, 1)$, so for large k

$$\|D\tilde{T}^t(\tilde{m})|_{\tilde{X}^s_{\tilde{m}}}\| < \lambda_1.$$

With the help of Lemma 2.2, we can prove the second part of (12.9) in the same fashion. Restricting our attention to $\tilde{W}^{cu}$ and considering the inverse of $\tilde{T}^t$ on it, (12.8) follows. $\square$

Proof of lemma 12.8.
  Let $\tilde{M}_\epsilon = \{x \in X \mid d(x, \tilde{M}) < \epsilon\}$.
  First, we claim that for any fixed $t > 0$, there exists $\epsilon > 0$ such that $\mathrm{Lip}\tilde{T}^t|_{\tilde{M}_\epsilon} < \infty$. In fact, there exists $a < \infty$ such that $\|D\tilde{T}^t\|_{\tilde{M}} = a$, since $\tilde{M}$ is compact and $D\tilde{T}^t(\cdot)$ is continuous. For each $\tilde{m}$, there exists $\epsilon_{\tilde{m}} > 0$ such that $\|D\tilde{T}^t\|_{B(\tilde{m}, \epsilon_{\tilde{m}})} < a + 1$. Choose a finite number of $\tilde{m}_i \in \tilde{M}$ such that $\tilde{M} \subset \cup_{i=1}^k B(\tilde{m}_i, \frac{\epsilon_i}{5(1+\bar{C})})$, where $\bar{C}$ is a constant satisfying

$$|\tilde{m} - \tilde{m}_0| \le d(\tilde{m}, \tilde{m}_0) \le \bar{C}|\tilde{m} - \tilde{m}_0|$$

for all $\tilde{m}, \tilde{m}_0 \in \tilde{M}$, where $d(\,\cdot\,)$ is defined in the proof of Lemma 2.2 (but on $\tilde{M}$), and $\epsilon_i = \epsilon_{\tilde{m}_i}$
  For $x_1 \in B(\tilde{m}_i, \frac{\epsilon_i}{5(1+\bar{C})}), x_2 \in B(\tilde{m}_j, \frac{\epsilon_j}{5(1+\bar{C})})$, we have

$$|\tilde{T}^t(x_1) - \tilde{T}^t(x_2)| \le (a+1)(\bar{C}|\tilde{m}_i - \tilde{m}_j| + \frac{\epsilon_i + \epsilon_j}{5(1+\bar{C})}). \tag{12.11}$$

If $\frac{\epsilon_i + \epsilon_j}{5(1+\bar{C})} + \bar{C}|\tilde{m}_i - \tilde{m}_j| \le (1+\bar{C})|x_1 - x_2|$ then we have

$$|\tilde{T}^t(x_1) - \tilde{T}^t(x_2)| \le (a+1)(1+\bar{C})|x_1 - x_2|. \tag{12.12}$$

Otherwise, without loss of generality, we may assume $\epsilon_i \ge \epsilon_j$ and we have

$$(1+\bar{C})|x_1 - x_2| < \frac{\epsilon_i + \epsilon_j}{5(1+\bar{C})} + \bar{C}|\tilde{m}_i - \tilde{m}_j| \le \bar{C}|x_1 - x_2| + \frac{\epsilon_i + \epsilon_j}{5},$$

which yields

$$|x_1 - x_2| \le \frac{2\epsilon_i}{5}.$$

Thus, $x_1, x_2 \in B(\tilde{m}_i, \epsilon_i)$. Hence,

$$|\tilde{T}^t(x_1) - \tilde{T}^t(x_2)| \leq (a+1)|x_1 - x_2|. \tag{12.13}$$

Let $\epsilon > 0$ be so small that $\tilde{M}_\epsilon \subset \cup_{i=1}^k B(\tilde{m}_i, \frac{\epsilon_i}{5(1+\bar{C})})$, therefore $\mathrm{Lip}\tilde{T}^t|_{\tilde{M}_\epsilon} \leq (a+1)(1+\bar{C})$. This completes the proof of the claim.

Let $E_{a,\epsilon} = \{t \in [0,1] : \mathrm{Lip}\tilde{T}^t|_{\tilde{M}_\epsilon} \leq a\}$, where a is a positive integer and $\epsilon \in \{\frac{1}{\ell} \mid \ell$ is a positive integer$\}$. It is clear that $E_{a,\epsilon}$ is closed and $[0,1] = \cup E_{a,\epsilon}$. The Baire Category Theorem implies that there exist a, ϵ and $t'' > t'$ such that $[t', t''] \subset E_{a,\epsilon}$, which implies that for all $t \in [t', t''], \|D\tilde{T}^t\|_{\tilde{M}} \leq a$. The same technique as in the proof of Lemma 2.3 gives the desired result. $\quad\square$

Summarizing all theorems and propositions, we obtain Theorems A′—D′.

REFERENCES

[An]   D. Anosov, *Geodesic flows on closed Riemann manifolds with negative curvature*, Proc. of the Stecklov Inst. of Math. **90** (1967), English translation, AMS, Providence, R.I., 1969.

[AG]   B. Aulbach and B. M. Garay, *Partial linearization for noninvertible mappings*, Z. Angew. Math. Phys. **45** (1994), 505-542.

[Ba]   J. M. Ball, *Saddle point analysis for an ordinary differential equation in a Banach space and an application to dynamic buckling of a beam*, Nonlinear Elasticity (R. W. Dickey, ed.), Academic Press, New York, 1973, pp. 93–160.

[BJ]   P. W. Bates and C. K. R. T. Jones, *Invariant manifolds for semilinear partial differential equations*, Dynamics Reported **2** (1989), 1–38, Wiley.

[BL]   P. W. Bates and K. Lu, *A Hartman-Grobman theorem for the Cahn-Hilliard and phase-field equations*, J. Dynamics and Differential Equations **6** (1994), 101–145.

[BLZ1] P. W. Bates, K. Lu and C. Zeng, *Invariant foliations near a normally hyperbolic invariant manifold for semiflow*, preprint (1996).

[BLZ2] P. W. Bates, K. Lu and C. Zeng, in preparation.

[BM]   N. Bogoliubov and Y. Mitropolsky, Asymptotic methods in the theory of nonlinear oscillations, Gordon and Breach, New York, 1961.

[BK]   I. U. Bronstein and A. Ya. Kopanskii, Smooth Invariant Manifolds and Normal Forms, vol. 7, World Scientific Series on Nonlinear Science, Singapore, 1994.

[BDL]   A. Burchard, B. Deng and K. Lu,, *Smooth Conjugacy of Center Manifolds*, Proceedings of the Royal Society of Edingburgh **120A** (1992), 61-77.

[Ca]   J. Carr, Applications of centre manifold theory, Springer-Verlag, New York, 1981.

[CLL]   S. Chow, X.. Lin and K. Lu, *Smooth invariant foliations in infinite dimensional spaces*, J. Differential Equations **94** (1991), 266–291.

[CL1]   S.-N. Chow and K. Lu, *Invariant manifolds for flows in Banach spaces*, J. Differential Equations **74** (1988), 285–317.

[CL2]   S-N. Chow and K. Lu, *Invariant Manifolds and foliations for Quasiperiodic Systems*, J. Differential Equations **117** (1995), 1-27.

[CFNT] P. Constantin, C. Foias, B. Nicolaenko, R. Témam, *Integral Manifolds and Inertial Manifolds for Dissipative Partial Differential Equations*, Appl. Math. Sciences, No. 70, Springer Verlag, New York, 1989.

[D]   B. Deng, *The existence of infinitely many traveling front and back waves in the Fitzhugh-Nagumo equations*, SIAM J. Math. Anal. **22** (1991), 1631-1650.

[Di]   S. P. Diliberto, *Perturbation theorems for periodic surfaces, I*, Rend. Circ. Mat. Palermo Ser. 2, **9** (1960), 265–299.

[DL]   G. Da Prato and A. Lunardi, *Stability, instability and center manifold theorems for fully nonlinear autonomous parabolic equations in Banach space*, Arch. Rat. Mech. Anal. **101** (1988), 115-141.

[F1]   N. Fenichel, *Persistence and smoothness of invariant manifolds for flows*, Indiana Univ. Math. Journal **21** (1971), 193–226.

[F2]   N. Fenichel, *Asymptotic stability with rate conditions*, Indiana Univ. Math. Journal **23** (1974), 1109–1137.

[F3]   N. Fenichel, *Asymptotic stability with rate conditions II*, Indiana Univ. Math. Journal **26** (1977), 81–93.

[F4]   N. Fenichel, *Geometric singular perturbation theory for ordinary differential equations*, J. Differential Equations **31** (1979), 53–98.

[FST]   C. Foias, G. R. Sell and R.Témam, *Inertial manifolds for nonlinear evolutionary equations*, J. Differential Equations **73** (1988), 309–353.

[G]   R. A. Gardner, *An invariant manifold analysis of electrophoretic traveling waves*, J. Dynamics and Differential Equations **5** (1993), 599-606.

[GS]   I. Gasser and P. Szmolyan, *A geometric singular perturbation analysis of detonation and deflagration waves*, SIAM J. Math. Anal. **24** (1993), 968-986.

[Ha]   J. Hadamard, *Sur l'iteration et les solutions asymptotiques des equations differentielles*, Bull. Soc. Math. France **29** (1901), 224–228.

[Ha1]   J. K. Hale, *Integral manifolds of perturbed differential systems*, Annals of Math. **73** (1961), 496–531.

[Ha2]   J. K. Hale, *On the method of Krylov-Bogoliubov-Metropolsky for the existence of integral manifolds*, Bol. Soc. Mex. (1960).

[Ha3]   J. K. Hale, *Ordinary Differential Equations* (1969), John Wiley, New York.

[HW]   G. Haller and S. Wiggins, *N-pulse homoclinic orbits in perturbations of resonant Hamiltonian systems*, Arch. Rat. Mech. Anal. **130** (1995), 25-101.

[Har]   P. Hartman,, *Ordinary Differential Equations* (1964), Wiley, New York.

[He]   D. Henry, *Geometric theory of semilinear parabolic equations*, Lecture Notes in Mathematics, vol. 840, Springer-Verlag, New York, 1981.

[HP]   M. W. Hirsch, and C. C. Pugh, *Stable manifolds and hyperbolic sets*, Global Analysis, Proc. Sympos. Pure Math. **14** (1970), Amer. Math. Soc., Providence, 133-163.

[HPS1] M. W. Hirsch, C. C. Pugh and M. Shub, *Invariant manifolds*, Bull. AMS **76** (1970), 1015–1019.

[HPS]   M. W. Hirsch, C. C. Pugh and M. Shub, *Invariant manifolds*, Lecture Notes in Mathematics, vol. 583, Springer-Verlag, New York, 1977.

[Hu]   G. Hufford, *Banach spaces and the perturbation of ordinary differential equations*, Contributions to the theory of nonlinear oscillations, Annals of Math. Studies, III Number 36, Princeton University Press, Princeton, 1956, pp. 173–195.

[Ir]   M. C. Irwin, *On the stable manifold theorem*, Bull. London Math. Soc. **2** (1970), 196.

[Jo]   C. K. R. T. Jones, *Geometric singular perturbation theory*, preprint, C.I.M.E. Lectures (1994).

[JK]   C. K. R. T. Jones and N. Kopell, *Tracking invariant manifolds with differential forms in singularly perturbed systems*, J. Differential Equations **108** (1994), 64-88.

[Ke]   A. Kelley, *The stable, center-stable, center, center-unstable, and unstable manifolds*, J. Differential Equations **3** (1967), 546-570.

[KP]   U. Kirchgraber and K. J. Palmer, *Geometry in the neighborhood of invariant manifolds of maps and flows and linearization*, Pitman Research Notes in Mathematics Series, Longman Scientific and Technical, 1990, published in the United States by John Wiley and Sons, Inc. New York.

[KW]   G. Kovacic and S. Wiggins, *Orbits homoclinic to resonances with applications to chaos in a model of the forced and damped sine-Gordon equation*, Physica-D **57** (1992), 185-225.

[KB]   N. Krylov and N. Bogoliubov, The application of methods of nonlinear mechanics to the theory of stationary oscillations, Publication 8 of the Ukrainian Academy of Science, Kiev, 1934.

[Ku]   J. Kurzweil, *Invariant manifolds for flows*, Proc. Sym. Diff. Equations and Dynamical Systems, Academic Press, New York, 1967.

[Ky1]   W. T. Kyner, *A fixed point theorem*, Ann. of Math. Studies **III** (1956), 197–205.

[Ky2]   W. T. Kyner, *Invariant manifolds*, Rend. Circ. Mat. Palermo, Ser. 2, **10** (1961), 98–110.

[Le]   N. Levinson, *Small periodic perturbations of an autonomous system with a stable orbit*, Annals of Math. (2) **52** (1950), 727–738.

[Li]   X-B. Lin, *Shadowing lemma and singularly perturbed boundary value problems*, SIAM J. Appl. Math. **49** (1989), 26-54.

[LMSW] Y. Li, D. W. McLaughlin, J. Shatah and S. Wiggins, *Persistent homoclinic orbits for perturbed nonlinear Schrödinger equations*, preprint (July 1995).

[LMW]   Y. Li, D. W. McLaughlin and S. Wiggins, *Invariant manifolds and their fibrations for perturbed NLS PDEs; graph transform approach*, in preparation (1995).

[Lu]   K. Lu, *A Hartman-Grobman theorem for scalar reaction-diffusion equations*, J. Differential Equations **93** (1991), 364-394.

[Ly]   A. M. Liapunov, *Problème géneral de la stabilité du mouvement*, Annals Math. Studies **17** (1947), Princeton, N.J. (originally published in Russian, 1892).

[Mn1]   R. Mañé, *Persistent manifolds are normally hyperbolic*, Trans. Amer. Math. Soc. **246** (1978), 261-283.

[Mn2]   R. Mañé, *Liapunov exponents and stable manifolds for compact transformations*, Geometrical Dynamics, Lecture Notes in Math., vol. 1007, Springer Verlag, New York, 1985, pp. 522–577.

[Ma]   M. Marcus, *Invariant surface theorem for a non-degenerate system*, Contribution to the theory of nonlinear oscillations, Annals of Math. Studies, Number 36, vol. III, Princeton University Press, Princeton, 1956, pp. 243–256.

[Mc]   J. McCarthy, *Stability of invariant manifolds*, abstract, Bull. Amer. Math. Soc. **61** (1955), 149–150.

[Mi]   A. Mielke, *Hamiltonian and Lagrangian flows on center manifolds*, Lecture Notes in Mathematics, vol. 1489, Springer-Verlag, New York, 1991.

[MPS]   J. Mallet-Paret and G. R. Sell, *Inertial manifolds for reaction diffusion equations in higher space dimensions*, J. Amer. Math. Soc. **1** (1988), 805–866.

[MS]   J.E. Marsden and J. Scheurle, *The construction and smoothness of invariant manifolds by the deformation method*, SIAM J. Math. Anal. **18** (1987), 1261-1274.

[Ny]   Y. Nyemark, *Integral manifolds for differential equations,*, Izv. Visch. Uchebn. Zabed. Radiophisica **X** (1967), 321-334.

[PT]   Palis and F. Takens, *Topological equivalence of normally hyperbolic dynamical systems*, Topology **16** (1977), 335–345.

[P1]   O. Perron, *Über Stabilität und asymptotisches verhalten der integrale von differentialgleichungssystemen*, Math. Z. **29** (1928), 129–160.

[P2]   O. Perron, *Über stabilität und asymptotisches Verhalten der lösungen eines systems endlicher Differenzengleichungen*, J. Reine Angew. Math. **161** (1929), 41–64.

[P3]   O. Perron, *Die stabilitätsfrage bei differentialgleichungen*, Math. Z. **32** (1930), 703–728.

[Pl]   V. A. Pliss, *A reduction principle in the theory of the stability of motion*, Izv. Akad. Nauk SSSR, Mat. Ser. **28** (1964), 1297-1324.

[PS]   V. A. Pliss and G. R. Sell, *Perturbations of attractors of differential equations*, J. Differential Equations **92** (1991), 100-124.

[Ru]   D. Ruelle, *Characteristic exponents and invariant manifolds in Hilbert space*, Annals of Math **115** (1982), 243–290.

[Sa]   R. Sacker, *A perturbation theorem for invariant manifolds and holder conti-
        nuity*, J. Math. Mech. **18** (1969), 705–762.

[SS]   R. Sacker and G. R. Sell, *Dichotomies for linear evolutionary equations in
        Banach spaces*, J. Differential Equations **113** (1994), 17-67.

[Se]   G. R. Sell, Private communication.

[Si]   Sijbrand, *Properties of center manifolds*, Trans. Amer. Math. Soc. **289**
        (1985), 431–469.

[Sm]   S. Smale, *Differentiable dynamical systems*, Bull. Amer. Math. Soc. **73**
        (1967), 747-817.

[Sz]   P. Szmolyan, *Transversal heterclinic and homoclinic orbits in singular per-
        turbed problems*, J. Differential Equations **92** (1991), 252-281.

[Te]   D. Terman, *The transition from bursting to continuous spiking in excitable
        membrane models*, J. Nonlinear Sci. **2** (1992), 135-182.

[Va]   A. Vanderbauwhede, *Center manifolds, normal forms and elementary bifur-
        cations*, Dynamics Reported **2** (1989), 89-169.

[VI]   A. Vanderbauwhede and G. Iooss, *Center manifold theory in infinite dimen-
        sions*, Dynamics Reported, Universite de Nice (1990), Preprint.

[VV]   S. Van Gils and A. Vanderbauwhede, *Center manifolds and contractions on
        a scale of Banach spaces*, J. Funct. Anal **72** (1987), 209–224.

[Wa]   C. E. Wayne, *Invariant manifolds for parabolic partial differential equations
        on unbounded domains*, Arch. Rat. Mech. Anal.

[W]   S. Wiggins, *Normally hyperbolic invariant manifolds in dynamical systems*,
        Springer-Verlag, New York, 1994.

[Yi]   Y. Yi, *A generalized integral manifold theorem*, J. Differential Equations **102**
        (1993), 153-187.

Department of Mathemtics
Brigham Young University
Provo, UT 84602

E-mail: peter@math.byu.edu, klu@math.byu.edu, zeng@math.byu.edu

Editorial Information

To be published in the *Memoirs*, a paper must be correct, new, nontrivial, and significant. Further, it must be well written and of interest to a substantial number of mathematicians. Piecemeal results, such as an inconclusive step toward an unproved major theorem or a minor variation on a known result, are in general not acceptable for publication. *Transactions* Editors shall solicit and encourage publication of worthy papers. Papers appearing in *Memoirs* are generally longer than those appearing in *Transactions* with which it shares an editorial committee.

As of June 30, 1998, the backlog for this journal was approximately 9 volumes. This estimate is the result of dividing the number of manuscripts for this journal in the Providence office that have not yet gone to the printer on the above date by the average number of monographs per volume over the previous twelve months, reduced by the number of issues published in four months (the time necessary for preparing an issue for the printer). (There are 6 volumes per year, each containing at least 4 numbers.)

A Copyright Transfer Agreement is required before a paper will be published in this journal. By submitting a paper to this journal, authors certify that the manuscript has not been submitted to nor is it under consideration for publication by another journal, conference proceedings, or similar publication.

Information for Authors and Editors

Memoirs are printed by photo-offset from camera copy fully prepared by the author. This means that the finished book will look exactly like the copy submitted.

The paper must contain a *descriptive title* and an *abstract* that summarizes the article in language suitable for workers in the general field (algebra, analysis, etc.). The *descriptive title* should be short, but informative; useless or vague phrases such as "some remarks about" or "concerning" should be avoided. The *abstract* should be at least one complete sentence, and at most 300 words. Included with the footnotes to the paper, there should be the 1991 *Mathematics Subject Classification* representing the primary and secondary subjects of the article. This may be followed by a list of *key words and phrases* describing the subject matter of the article and taken from it. A list of the numbers may be found in the annual index of *Mathematical Reviews*, published with the December issue starting in 1990, as well as from the electronic service e-MATH [**telnet e-MATH.ams.org** (or **telnet 130.44.1.100**). Login and password are **e-math**]. For journal abbreviations used in bibliographies, see the list of serials in the latest *Mathematical Reviews* annual index. When the manuscript is submitted, authors should supply the editor with electronic addresses if available. These will be printed after the postal address at the end of each article.

Electronically prepared papers. The AMS encourages submission of electronically prepared papers in $\mathcal{A}_{\mathcal{M}}\mathcal{S}$-TeX or $\mathcal{A}_{\mathcal{M}}\mathcal{S}$-LaTeX. The Society has prepared author packages for each AMS publication. Author packages include instructions for preparing electronic papers, the *AMS Author Handbook*, samples, and a style file that generates the particular design specifications of that publication series for both $\mathcal{A}_{\mathcal{M}}\mathcal{S}$-TeX and $\mathcal{A}_{\mathcal{M}}\mathcal{S}$-LaTeX.

Authors with FTP access may retrieve an author package from the Society's Internet node **e-MATH.ams.org** (130.44.1.100). For those without FTP access, the author package can be obtained free of charge by sending e-mail to **pub@ams.org** (Internet) or from the Publication Division, American Mathematical Society, P.O. Box 6248, Providence, RI 02940-6248. When requesting an author package, please specify $\mathcal{AMS}$-TEX or $\mathcal{AMS}$-LATEX, Macintosh or IBM (3.5) format, and the publication in which your paper will appear. Please be sure to include your complete mailing address.

Submission of electronic files. At the time of submission, the source file(s) should be sent to the Providence office (this includes any TEX source file, any graphics files, and the DVI or PostScript file).

Before sending the source file, be sure you have proofread your paper carefully. The files you send must be the EXACT files used to generate the proof copy that was accepted for publication. For all publications, authors are required to send a printed copy of their paper, which exactly matches the copy approved for publication, along with any graphics that will appear in the paper.

TEX files may be submitted by email, FTP, or on diskette. The DVI file(s) and PostScript files should be submitted only by FTP or on diskette unless they are encoded properly to submit through e-mail. (DVI files are binary and PostScript files tend to be very large.)

Files sent by electronic mail should be addressed to the Internet address **pub-submit@ams.org**. The subject line of the message should include the publication code to identify it as a Memoir. TEX source files, DVI files, and PostScript files can be transferred over the Internet by FTP to the Internet node **e-math.ams.org** (130.44.1.100).

Electronic graphics. Figures may be submitted to the AMS in an electronic format. The AMS recommends that graphics created electronically be saved in Encapsulated PostScript (EPS) format. This includes graphics originated via a graphics application as well as scanned photographs or other computer-generated images.

If the graphics package used does not support EPS output, the graphics file should be saved in one of the standard graphics formats—such as TIFF, PICT, GIF, etc.—rather than in an application-dependent format. Graphics files submitted in an application-dependent format are not likely to be used. No matter what method was used to produce the graphic, it is necessary to provide a paper copy to the AMS.

Authors using graphics packages for the creation of electronic art should also avoid the use of any lines thinner than 0.5 points in width. Many graphics packages allow the user to specify a "hairline" for a very thin line. Hairlines often look acceptable when proofed on a typical laser printer. However, when produced on a high-resolution laser imagesetter, hairlines become nearly invisible and will be lost entirely in the final printing process.

Screens should be set to values between 15% and 85%. Screens which fall outside of this range are too light or too dark to print correctly.

Any inquiries concerning a paper that has been accepted for publication should be sent directly to the Editorial Department, American Mathematical Society, P. O. Box 6248, Providence, RI 02940-6248.

Selected Titles in This Series

(*Continued from the front of this publication*)

(See the AMS catalog for earlier titles)